Everyday Mathematics®

The University of Chicago School Mathematics Project

Student Math Journal
Volume 1

Grade 2

The University of Chicago School Mathematics Project (UCSMP)

Max Bell, Director, UCSMP Elementary Materials Component; Director, *Everyday Mathematics* First Edition
James McBride, Director, *Everyday Mathematics* Second Edition
Andy Isaacs, Director, *Everyday Mathematics* Third Edition
Amy Dillard, Associate Director, *Everyday Mathematics* Third Edition

Authors

Max Bell
Jean Bell
John Bretzlauf
Amy Dillard
Robert Hartfield

Andy Isaacs
James McBride
Cheryl G. Moran*
Kathleen Pitvorec
Peter Saecker

Third Edition only

Technical Art
Diana Barrie

Teachers in Residence
Kathleen Clark, Patti Satz

Editorial Assistant
John Wray

Contributors

Robert Balfanz, Judith Busse, Mary Ellen Dairyko, Lynn Evans, James Flanders, Dorothy Freedman, Nancy Guile Goodsell, Pam Guastafeste, Nancy Hanvey, Murray Hozinsky, Deborah Arron Leslie, Sue Lindsley, Mariana Mardrus, Carol Montag, Elizabeth Moore, Kate Morrison, William D. Pattison, Joan Pederson, Brenda Penix, June Ploen, Herb Price, Dannette Riehle, Ellen Ryan, Marie Schilling, Susan Sherrill, Patricia Smith, Robert Strang, Jaronda Strong, Kevin Sweeney, Sally Vongsathorn, Esther Weiss, Francine Williams, Michael Wilson, Izaak Wirzup

Photo Credits

Getty Images, cover, *right,* p. vii *top;* ©Ingram Publishing/Alamy, p. 93 *top row center, bottom row second from left;* ©Linda Lewis; Frank Lane Picture Agency/CORBIS, cover, *bottom left;* Getty Images, p. viii; ©Star/zefa/Corbis, cover, *center;* ©Stockbyte, p. vii *bottom;* ©Michael Tupy/iStock, Inc., p. v *top.*

---

www.WrightGroup.com

Copyright © 2007 by Wright Group/McGraw-Hill.

All rights reserved. Except as permitted under the United States Copyright Act, no part of this publication may be reproduced or distributed in any form by any means, or stored in a database or retrieval system, without the prior written permission of the publisher, unless otherwise indicated.

Printed in the United States of America.

Send all inquiries to:
Wright Group/McGraw-Hill
P.O. Box 812960
Chicago, IL 60681

ISBN 0-07-604554-4

16 CPC 12 11 10

Contents

UNIT 1 — Numbers and Routines

Number Sequences	1
Coins	2
Calendar for the Month	3
Time	4
Addition Facts	5
Counting Bills	6
Math Boxes 1◆6	7
Math Boxes 1◆7	8
Number-Grid Puzzles	9
Math Boxes 1◆8	10
Broken Calculator	11
Hundreds-Tens-and-Ones Problems	12
Math Boxes 1◆9	13
Counting with a Calculator	14
Broken Calculator	15
Math Boxes 1◆10	16
Using <, >, and =	17
Math Boxes 1◆11	18
Math Boxes 1◆12	19
Math Boxes 1◆13	20

Contents **iii**

UNIT 2 Addition and Subtraction Facts

Number Stories	21
Number-Grid Puzzles	22
Math Boxes 2•1	23
Facts Power Table	24
Distances on a Number Grid	25
Math Boxes 2•2	26
Addition/Subtraction Facts Table	27
Domino-Dot Patterns	28
Doubles or Nothing Record Sheet	29
Math Boxes 2•3	30
+9 Facts	31
Math Boxes 2•4	32
Addition Facts	33
Math Boxes 2•5	34
Domino Facts	35
Math Boxes 2•6	36
Subtraction Number Stories	37
Math Boxes 2•7	38
Using a Pan Balance and a Spring Scale	39
Math Boxes 2•8	40
Name-Collection Boxes	41
Pan-Balance Problems	42
Math Boxes 2•9	44
Frames-and-Arrows Problems	45
Math Boxes 2•10	46
"What's My Rule?"	47
Math Boxes 2•11	48
Math Boxes 2•12	49
Subtract 9 or 8	50
Math Boxes 2•13	51
Math Boxes 2•14	52

iv Contents

UNIT 3   Place Value, Money, and Time

Place Value 53
Math Boxes 3◆1 54
Spinning for Money 55
Fruit and Vegetables Stand Poster 56
Buying Fruit and Vegetables 57
Math Boxes 3◆2 58
What Time Is It? 59
Math Boxes 3◆3 60
Build a Number 61
Geoboard Dot Paper (7 × 7) 62
Magic Squares 63
Math Boxes 3◆4 64
Dollar Rummy 65
Pockets Data Table 66
Graphing Pockets Data 67
Math Boxes 3◆5 68
Two-Rule Frames and Arrows 69
Reading a Graph 71
Math Boxes 3◆6 72
Shopping at the Fruit and Vegetables Stand . . . 73
Math Boxes 3◆7 74
Making Change 75
Buying from a Vending Machine 76
Math Boxes 3◆8 78
Math Boxes 3◆9 79

Contents   **v**

UNIT 4 Addition and Subtraction

Fish Poster	80
"Fishy" Stories	81
Distances on a Number Grid	83
Math Boxes 4♦1	84
Parts-and-Total Number Stories	85
Math Boxes 4♦2	86
Temperatures	87
Math Boxes 4♦3	88
Parts-and-Total Number Stories	89
Temperature Changes	90
Math Boxes 4♦4	92
School Supply Store	93
Math Boxes 4♦5	94
Shopping Activity	95
Shopping Poster	96
Shopping Problems	97
Math Boxes 4♦6	98
Measuring Lengths with a Tape Measure	99
Tiling Surfaces with Shapes	101
An Attribute Rule	103
Math Boxes 4♦7	104
Addition Practice	105
Math Boxes 4♦8	106
Partial-Sums Addition Using Base-10 Blocks	107
Addition Practice	108
The Time of Day	109
Math Boxes 4♦9	110
Math Boxes 4♦10	111

vi Contents

UNIT 5 3-D and 2-D Shapes

Math Boxes 5•1	112
Using Secret Codes to Draw Shapes	113
Math Boxes 5•2	114
Parallel or Not Parallel?	115
Parallel Line Segments	116
"What's My Rule?" and Frames and Arrows	118
Math Boxes 5•3	119
Polygons	120
Math Boxes 5•4	121
Math Boxes 5•5	122
Connecting Points	123
3-D Shapes Poster	124
What's the Shape?	125
Math Boxes 5•6	126
Math Boxes 5•7	127
Symmetrical Shapes	128
Math Boxes 5•8	129
Math Boxes 5•9	130

UNIT 6 Whole-Number Operations and Number Stories

Three Addends Directions 131
Ballpark Estimates 132
Math Boxes 6•1 133
Comparison Number Stories 134
Math Boxes 6•2 136
What Is Your Favorite Food? 137
Comparing Fish 138
Math Boxes 6•3 140
Addition and Subtraction Number Stories 141
Ballpark Estimates 142
Math Boxes 6•4 143
Subtraction 144
Math Boxes 6•5 145
How Many Children Get *n* Things? 146
Math Boxes 6•6 147
Multiplication Stories 148
Math Boxes 6•7 149
Array Number Stories 150
Multiplication Number Stories 151
Math Boxes 6•8 152
Math Boxes 6•9 153
Array Bingo Directions 154
Division Problems 156
Math Boxes 6•10 157
Math Boxes 6•11 158
Table of Equivalencies 159

Activity Sheets

+,− Fact Triangles 1 .... **Activity Sheet 1**
+,− Fact Triangles 2 .... **Activity Sheet 2**
+,− Fact Triangles 3 .... **Activity Sheet 3**
+,− Fact Triangles 4 .... **Activity Sheet 4**

viii   Contents

LESSON 1·1 Number Sequences

Fill in the missing numbers.

1. 21, ___, 24, ___, ___, ___, 28, ___

2. 49, ___, ___, ___, 54, ___, ___, 57

3. 127, ___, ___, 130, ___, ___, ___, 134, ___, ___

4. 249, ___, ___, ___, ___, 254, ___, ___, 257, ___

5. 599, ___, ___, ___, 603, ___, 606, ___, ___

6. 1,027, 1,028, ___, ___, ___, ___, 1,032, 1,033, ___

one 1

LESSON 1·2 Coins

1.   = _____ ¢

2.   = _____ ¢

3.   = _____ ¢

4.   = _____ ¢

5.   = _____ ¢

6.   = _____ ¢

2 two

Date _____ Time _____

LESSON 1·3 **Calendar for the Month**

Month _____

	Sunday	Monday	Tuesday	Wednesday	Thursday	Friday	Saturday

three 3

Date　　　　　　　　　　　　　　　Time

LESSON 1·3　Time

Write the time.

1.

2.

3.

____ : ____　　　____ : ____　　　____ : ____

Draw the hands.

4.

5.

6.

5:00　　　　　　9:30　　　　　　1:45

Draw the hands and write the time.

7.

8.

9.

____ : ____　　　____ : ____　　　____ : ____

4　four

Date Time

LESSON 1·4 Addition Facts

1.

   2 + 4 = ____

2.

   5 + 5 = ____

3. 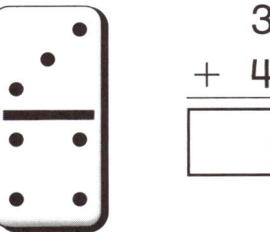 $\begin{array}{r} 3 \\ + 4 \\ \hline \square \end{array}$

4. $\begin{array}{r} 7 \\ + 9 \\ \hline \square \end{array}$

5. 6 + 8 = ____   6. 8 + ____ = 11   7. ____ + 4 = 10

Addition Top-It

Write a number model for each player's cards.

Then write <, >, or = in the box.

_____ = _____   □   _____ = _____

_____ = _____   □   _____ = _____

_____ = _____   □   _____ = _____

| Date | Time |

Counting Bills

Write the amount.

1.   = $_____

2.   = $_____

3.   = $_____

4.   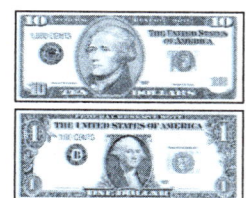   = $_____

Try This

5.   = $_____

6 six

Date _____ Time _____

LESSON 1·6 Math Boxes

1. Fill in the missing numbers.

   a. 36, _____, 38, _____

   b. _____, 12, 13, _____

   c. 89, _____, 91, _____

   d. _____, 147, _____, 149, _____

2. Circle the tens digit.

   437

   Circle the ones digit.

   18

   Circle the hundreds digit.

   206

3. How likely is it that our class will go on a field trip today? Circle.

   certain

   likely

   unlikely

   impossible

4. Today is

   _____ _____, _____.
   (month)   (day)   (year)

   The date 1 week from today will be _____.

5. Fill in the circle next to the name of the shape.

   Ⓐ triangle

   Ⓑ rectangle

   Ⓒ pentagon

6. Write two even and two odd numbers.

   even _____

   even _____

   odd _____

   odd _____

seven **7**

Date _____ Time _____

LESSON 1·7 Math Boxes

1. How much money?

   _____ ¢

2. Fill in the missing numbers.

	4			7
13		15		

3. Show 16 with tally marks.

   _____

4. Write the time.

   _____ : _____

5. Solve.

6. Fill in the missing frames.

     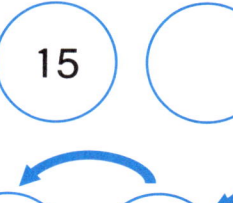

8 eight

Date _____ Time _____

LESSON 1·8 Number-Grid Puzzles

						80		
9								
7					66			
		35						
2								
		21				61		

nine 9

Date _____ Time _____

LESSON 1·8 Math Boxes

1. Fill in the missing numbers.

47 ____ ____ ____

2. Write the missing numbers.

124, _____

126, _____

128, _____

3. Show 2 ways to make 5 cents.

4. Write odd or even.

3 _____

8 _____

1 _____

27 _____

5. Fill in the missing frames.

Rule +10

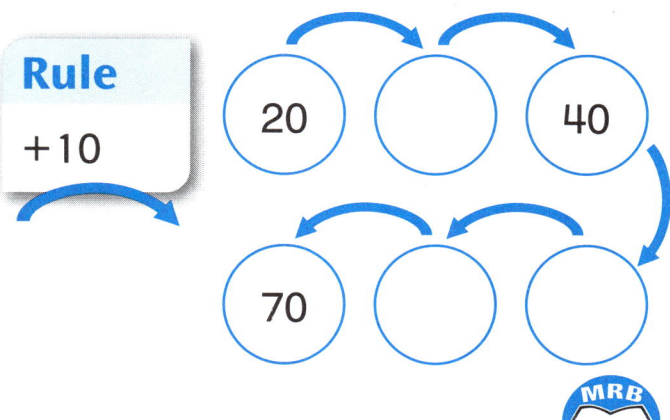

6. Solve.

Unit

____ = 6 + 5

7 + 5 = ____

____ = 8 + 5

9 + 5 = ____

10 ten

Date　　　　　　　　　　Time

LESSON 1·9　Broken Calculator

Example: Show 17.
Broken key is (7).
Show several ways:

11 + 6
20 − 3
8 + 8 + 1

1. Show 3.
   Broken key is (3).
   Show several ways:

   _____

   _____

   _____

2. Show 20.
   Broken key is (2).
   Show several ways:

   _____

   _____

   _____

3. Show 22.
   Broken key is (2).
   Show several ways:

   _____

   _____

   _____

4. Show 12.
   Broken key is (1).
   Show several ways:

   _____

   _____

   _____

5. Make up your own.
   Show _____.
   Broken key is _____.
   Show several ways:

   _____

   _____

   _____

eleven **11**

Date _____ Time _____

LESSON 1·9 Hundreds-Tens-and-Ones Problems

Solve. Use your tool-kit bills to help you.

Example:

2 [$10]   3 [$1]   How much? $ _____

1.

8 [$10]   4 [$1]   How much? $ _____

2.

5 [$100]   4 [$10]   3 [$1]   How much? $ _____

3.

7 [$100]   6 [$10]   9 [$1]   How much? $ _____

4.

8 [$100]   2 [$10]   0 [$1]   How much? $ _____

5.

3 [$100]   0 [$10]   4 [$1]   How much? $ _____

Try This

6.

22 [$100]   5 [$10]   7 [$1]   How much? $ _____

12 twelve

Date　　　　　　　　　　Time

LESSON 1·9　Math Boxes

1. How much money?

     D

_____ ¢

2. Fill in the missing numbers.

2			
	13		15

3. Show 23 with tally marks.

_____

4. What time is it?

_____ : _____

In $\frac{1}{2}$ hour, it will be _____ : _____.

5. Solve. Fill in the circle next to the best answer.

_____ + 6 = 8

Ⓐ 14
Ⓑ 2
Ⓒ 8
Ⓓ 10

6. Write the arrow rule. Fill in the missing frames.

Rule

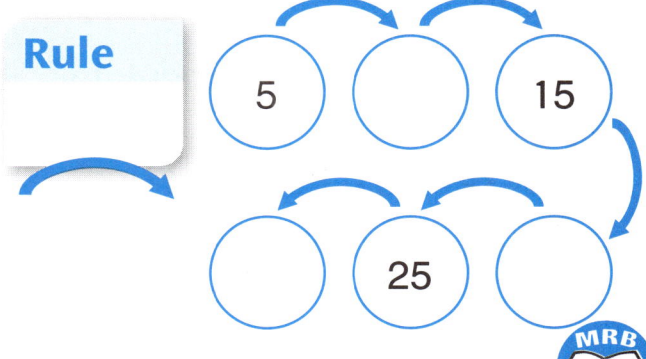

5　　　　　15

　　25

thirteen 13

Date _____ Time _____

LESSON 1·10 Counting with a Calculator

1. Count by 7s on your calculator. Write the numbers.

   7, _____, _____, _____, _____, _____, _____

   What number is added each time you press ⊜? _____

2. Count by 6s on your calculator. Write the numbers.
   Circle the 1s digit in each number.

   6, _____, _____, _____, _____, _____, _____, _____, _____, _____

   What number is added each time you press ⊜? _____

   What pattern do you see in the 1s digits? _____

3. Count by 4s on your calculator. Write the numbers.
   Circle the 1s digit in each number.

   4, _____, _____, _____, _____, _____, _____, _____, _____, _____

   What number is added each time you press ⊜? _____

   What pattern do you see in the 1s digits? _____

Try This

4. Jim counted on the calculator. He wrote these numbers: 3, 5, 7, 9, 11.

   What keys did he press? _____

14   fourteen

LESSON 1·10	**Broken Calculator**

1. Show 8.
   Broken key is ⑧.
   Show several ways:

   _____

   _____

   _____

2. Show 30.
   Broken key is ③.
   Show several ways:

   _____

   _____

   _____

3. Show 15.
   Broken key is ⑤.
   Show several ways:

   _____

   _____

   _____

4. Show 26.
   Broken key is ⑥.
   Show several ways:

   _____

   _____

   _____

5. Make up your own.
   Show _____.
   Broken key is _____.
   Show several ways:

   _____

   _____

   _____

6. Make up your own.
   Show _____.
   Broken key is _____.
   Show several ways:

   _____

   _____

   _____

fifteen **15**

Date   Time

LESSON 1·10 Math Boxes

1. Fill in the missing numbers.

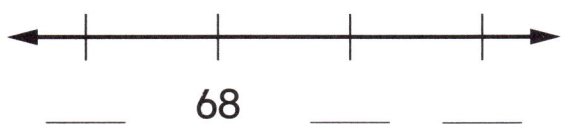
___ 68 ___ ___

2. Write these numbers in order. Start with the smallest number.
40   23   81

_____, _____, _____

3. Show two ways to make 35¢.

4. Write even or odd.

20 _____

5 _____

17 _____

33 _____

5. Fill in the missing frames.

Rule
+10

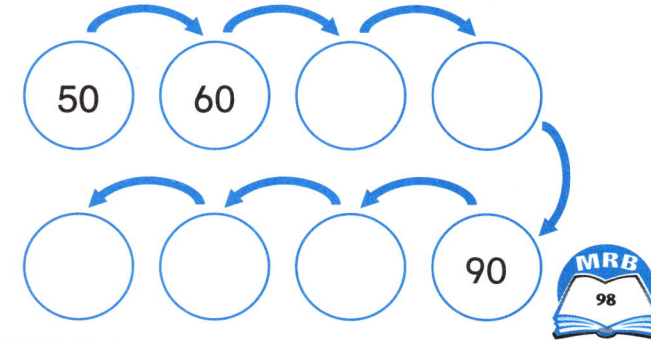

6. Solve.

```
  9     2     6
+ 5   + 5   + 5
 □     □     □

  8     5     5
+ 5   + 7   + 5
 □     □     □
```

Unit

16   sixteen

LESSON 1·11   Using <, >, and =

3 < 5
3 is less than 5.

5 > 3
5 is greater than 3.

Write <, >, or =.

1. 61 _____ 26
2. 18 _____ 81
3. 107 _____ 57
4. 114 _____ 114
5. 299 _____ 302
6. 1,032 _____ 1,132

Try This

7. 15 _____ 7 + 8
8. 9 + 2 _____ 4 + 5
9. 5 + 6 _____ 8 + 4

Write the total amounts. Then write <, >, or =.

Example: Ⓓ Ⓝ Ⓝ Ⓟ Ⓟ = _22_ ¢ _<_ _26_ ¢ = Ⓠ Ⓟ

10. Ⓝ Ⓝ Ⓓ Ⓟ = _____ ¢ _____ _____ ¢ = Ⓠ Ⓝ Ⓓ Ⓝ

11. Ⓝ Ⓓ Ⓟ Ⓠ = _____ ¢ _____ _____ ¢ = Ⓝ Ⓓ Ⓓ Ⓝ Ⓟ Ⓠ

12. Ⓓ Ⓓ Ⓠ Ⓓ = _____ ¢ _____ _____ ¢ = Ⓓ Ⓝ Ⓟ Ⓓ

seventeen 17

Date _____ Time _____

LESSON 1·11 Math Boxes

1. How much money? Circle the best answer.

    (D) (D) (D)
    (N) (P)

    A 41¢
    B 36¢
    C 82¢
    D 5¢

    _____ ¢

2. Fill in the missing numbers.

3. Write the number.

    |||| |||| |||| ||||
    |||| |||| |||

    _____

4. Draw hands to show 6:15.

5. Solve. **Unit**

    6 + 6 = _____

    9 + 9 = _____

    20 + 20 = _____

    _____ = 100 + 100

6. Solve. **Unit**

    139 + 1 = _____

    1 + 479 = _____

    _____ = 299 + 1

    _____ = 398 + 2

18 eighteen

Date   Time

LESSON 1·12 Math Boxes

1. Fill in the missing numbers.

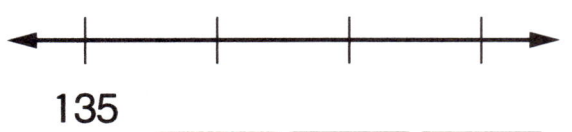

135 _____ _____ _____

2. Fill in the oval next to the numbers that are in order from the smallest to the largest.

◯ 103, 29, 86

◯ 29, 86, 103

◯ 29, 103, 86

◯ 86, 29, 103

3. Write the amount.

Ⓠ Ⓓ Ⓝ Ⓝ Ⓝ Ⓟ

_____ ¢

4. Write even or odd.

4 _____

7 _____

10 _____

46 _____

5. Write the arrow rule. Fill in the missing frames.

Rule

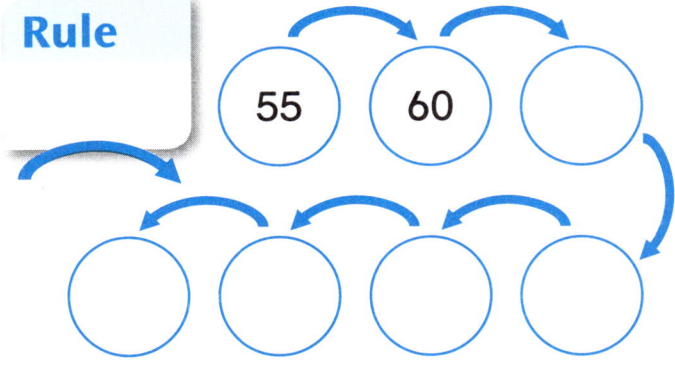

6. Fill in the blanks.

25, 35, _____, _____, _____

nineteen **19**

Date   Time

LESSON 1·13   Math Boxes

1. Write 6 names for 9.

     9

2. Solve.

     Unit

$4 + 5 =$ _____

$5 + 3 =$ _____

    8   11
$+ \ 5$ $+ \ 5$

3. Use a number grid. How many spaces from:

17 to 26? _____

49 to 65? _____

4. Solve.

     Unit

$14 + 1 =$ _____

$136 + 1 =$ _____

$291 + 1 =$ _____

$279 + 1 =$ _____

5. Fill in the missing frames.

Rule
+10

40   ◯   60

80   ◯

6. Solve.

$2 + \boxed{} = 6$

20   twenty

Date _____ Time _____

LESSON 2·1 **Number Stories**

Write an addition number story about what you see in the picture. Write a label in the unit box. Find the answer. Write a number model.

Example: 7 ducks in the water. 5 ducks in the grass. How many ducks in all?

Answer the question: __12 ducks__
                         (unit)

Number model: __7__ + __5__ = __12__

Unit
ducks

Story: _____

_____

_____

_____

_____

Unit

Answer the question: _____
                         (unit)

Number model: _____ + _____ = _____

twenty-one **21**

LESSON 2·1

Number-Grid Puzzles

Complete the number-grid puzzles.

20					
	28			58	
15					
	33				
				52	
11					

	349				
				378	
		357			
336					
				373	
332					
			361		

22 twenty-two

Date   Time

 LESSON 2·1 Math Boxes

1. Six apples are red. Five apples are green. How many apples in all?

   _____

   Unit
   apples

   Number Model

   _____

2. Use your calculator.

   Show 14.
   Broken key is ①.
   Show 2 ways:

   _____

   _____

3. Fill in the blanks.

   83, _____, 81, _____, _____, 78

4. Use < or >.

   4 + 5 _____ 10

   12 _____ 7 + 4

   15 _____ 8 + _____

   6 + 7 _____ 15 − 4

5. Write the time.

   _____ : _____ _____

6. How much money? Fill in the circle next to the best answer.

   Ⓐ $2.40   Ⓑ $11.45

   Ⓒ $11.40   Ⓓ $2.45

twenty-three **23**

LESSON 2·2 **Facts Power Table**

$\begin{array}{r}0\\+\,0\end{array}$	$\begin{array}{r}0\\+\,1\end{array}$	$\begin{array}{r}0\\+\,2\end{array}$	$\begin{array}{r}0\\+\,3\end{array}$	$\begin{array}{r}0\\+\,4\end{array}$	$\begin{array}{r}0\\+\,5\end{array}$	$\begin{array}{r}0\\+\,6\end{array}$	$\begin{array}{r}0\\+\,7\end{array}$	$\begin{array}{r}0\\+\,8\end{array}$	$\begin{array}{r}0\\+\,9\end{array}$
$\begin{array}{r}1\\+\,0\end{array}$	$\begin{array}{r}1\\+\,1\end{array}$	$\begin{array}{r}1\\+\,2\end{array}$	$\begin{array}{r}1\\+\,3\end{array}$	$\begin{array}{r}1\\+\,4\end{array}$	$\begin{array}{r}1\\+\,5\end{array}$	$\begin{array}{r}1\\+\,6\end{array}$	$\begin{array}{r}1\\+\,7\end{array}$	$\begin{array}{r}1\\+\,8\end{array}$	$\begin{array}{r}1\\+\,9\end{array}$
$\begin{array}{r}2\\+\,0\end{array}$	$\begin{array}{r}2\\+\,1\end{array}$	$\begin{array}{r}2\\+\,2\end{array}$	$\begin{array}{r}2\\+\,3\end{array}$	$\begin{array}{r}2\\+\,4\end{array}$	$\begin{array}{r}2\\+\,5\end{array}$	$\begin{array}{r}2\\+\,6\end{array}$	$\begin{array}{r}2\\+\,7\end{array}$	$\begin{array}{r}2\\+\,8\end{array}$	$\begin{array}{r}2\\+\,9\end{array}$
$\begin{array}{r}3\\+\,0\end{array}$	$\begin{array}{r}3\\+\,1\end{array}$	$\begin{array}{r}3\\+\,2\end{array}$	$\begin{array}{r}3\\+\,3\end{array}$	$\begin{array}{r}3\\+\,4\end{array}$	$\begin{array}{r}3\\+\,5\end{array}$	$\begin{array}{r}3\\+\,6\end{array}$	$\begin{array}{r}3\\+\,7\end{array}$	$\begin{array}{r}3\\+\,8\end{array}$	$\begin{array}{r}3\\+\,9\end{array}$
$\begin{array}{r}4\\+\,0\end{array}$	$\begin{array}{r}4\\+\,1\end{array}$	$\begin{array}{r}4\\+\,2\end{array}$	$\begin{array}{r}4\\+\,3\end{array}$	$\begin{array}{r}4\\+\,4\end{array}$	$\begin{array}{r}4\\+\,5\end{array}$	$\begin{array}{r}4\\+\,6\end{array}$	$\begin{array}{r}4\\+\,7\end{array}$	$\begin{array}{r}4\\+\,8\end{array}$	$\begin{array}{r}4\\+\,9\end{array}$
$\begin{array}{r}5\\+\,0\end{array}$	$\begin{array}{r}5\\+\,1\end{array}$	$\begin{array}{r}5\\+\,2\end{array}$	$\begin{array}{r}5\\+\,3\end{array}$	$\begin{array}{r}5\\+\,4\end{array}$	$\begin{array}{r}5\\+\,5\end{array}$	$\begin{array}{r}5\\+\,6\end{array}$	$\begin{array}{r}5\\+\,7\end{array}$	$\begin{array}{r}5\\+\,8\end{array}$	$\begin{array}{r}5\\+\,9\end{array}$
$\begin{array}{r}6\\+\,0\end{array}$	$\begin{array}{r}6\\+\,1\end{array}$	$\begin{array}{r}6\\+\,2\end{array}$	$\begin{array}{r}6\\+\,3\end{array}$	$\begin{array}{r}6\\+\,4\end{array}$	$\begin{array}{r}6\\+\,5\end{array}$	$\begin{array}{r}6\\+\,6\end{array}$	$\begin{array}{r}6\\+\,7\end{array}$	$\begin{array}{r}6\\+\,8\end{array}$	$\begin{array}{r}6\\+\,9\end{array}$
$\begin{array}{r}7\\+\,0\end{array}$	$\begin{array}{r}7\\+\,1\end{array}$	$\begin{array}{r}7\\+\,2\end{array}$	$\begin{array}{r}7\\+\,3\end{array}$	$\begin{array}{r}7\\+\,4\end{array}$	$\begin{array}{r}7\\+\,5\end{array}$	$\begin{array}{r}7\\+\,6\end{array}$	$\begin{array}{r}7\\+\,7\end{array}$	$\begin{array}{r}7\\+\,8\end{array}$	$\begin{array}{r}7\\+\,9\end{array}$
$\begin{array}{r}8\\+\,0\end{array}$	$\begin{array}{r}8\\+\,1\end{array}$	$\begin{array}{r}8\\+\,2\end{array}$	$\begin{array}{r}8\\+\,3\end{array}$	$\begin{array}{r}8\\+\,4\end{array}$	$\begin{array}{r}8\\+\,5\end{array}$	$\begin{array}{r}8\\+\,6\end{array}$	$\begin{array}{r}8\\+\,7\end{array}$	$\begin{array}{r}8\\+\,8\end{array}$	$\begin{array}{r}8\\+\,9\end{array}$
$\begin{array}{r}9\\+\,0\end{array}$	$\begin{array}{r}9\\+\,1\end{array}$	$\begin{array}{r}9\\+\,2\end{array}$	$\begin{array}{r}9\\+\,3\end{array}$	$\begin{array}{r}9\\+\,4\end{array}$	$\begin{array}{r}9\\+\,5\end{array}$	$\begin{array}{r}9\\+\,6\end{array}$	$\begin{array}{r}9\\+\,7\end{array}$	$\begin{array}{r}9\\+\,8\end{array}$	$\begin{array}{r}9\\+\,9\end{array}$

twenty-four

Date _____ Time _____

LESSON 2·2 Distances on a Number Grid

Example: How many spaces do you move to go from 17 to 23 on the number grid?

Solution: Place a marker on 17. You move the marker 6 spaces before landing on 23.

11	12	13	14	15	16	17	18	19	20
21	22	23	24	25	26	27	28	29	30

How many spaces from:

23 to 28? _____   15 to 55? _____   39 to 59? _____

27 to 42? _____   34 to 26? _____   54 to 42? _____

15 to 25? _____   26 to 34? _____

1	2	3	4	5	6	7	8	9	10
11	12	13	14	15	16	17	18	19	20
21	22	23	24	25	26	27	28	29	30
31	32	33	34	35	36	37	38	39	40
41	42	43	44	45	46	47	48	49	50
51	52	53	54	55	56	57	58	59	60

twenty-five

Date Time

LESSON 2·2 Math Boxes

1. Count by 3s.
   Use your calculator.

   ___, 6, ___, ___, 15, ___

2. Fill in the missing numbers.

   455, _____, _____, 458

3. Solve.

   4 + 3 = ___

   Unit

   10 − 7 = ___

   ```
     5        8
   + 4      − 3
   ```

4. Show $1.00 three ways.
   Use Q, D, and N.

5. Mrs. Satz's Class's Favorite Colors

Colors	Tallies
Red	‖‖‖‖ ‖‖‖‖
Blue	‖‖‖‖ ///
Green	‖‖‖‖
Yellow	//

   Which color is the most popular? _____

6. Fill in the circle that names the number.

   5 ones   Ⓐ 564
   6 hundreds   Ⓑ 356
   3 tens   Ⓒ 635
            Ⓓ 536

26 twenty-six

Date _____   Time _____

Addition/Subtraction Facts Table

+,−	0	1	2	3	4	5	6	7	8	9
0	0	1	2	3	4	5	6	7	8	9
1	1	2	3	4	5	6	7	8	9	10
2	2	3	4	5	6	7	8	9	10	11
3	3	4	5	6	7	8	9	10	11	12
4	4	5	6	7	8	9	10	11	12	13
5	5	6	7	8	9	10	11	12	13	14
6	6	7	8	9	10	11	12	13	14	15
7	7	8	9	10	11	12	13	14	15	16
8	8	9	10	11	12	13	14	15	16	17
9	9	10	11	12	13	14	15	16	17	18

twenty-seven

Date	Time

LESSON 2·3 Domino-Dot Patterns

Draw the missing dots on the dominoes. Find the total number on both halves.

1. double 2

2
+ 2

2. double 3

3
+ 3

3. double 4

4
+ 4

4. double 5

5
+ 5

5. double 6

6
+ 6

6. double 7

7
+ 7

7. double 8

8
+ 8

8. double 9

9
+ 9

Find the total number of dots.

9.

4
+ 5

10.

6
+ 5

11.

6
+ 7

12.

8
+ 7

28 twenty-eight

Date _____ Time _____

LESSON 2·3
Doubles or Nothing Record Sheet

Round 1

Total _____

Round 2

Total _____

Round 3

Total _____

Grand Total _____

twenty-nine **29**

Date _____ Time _____

LESSON 2·3 Math Boxes

1. Julie had 10 crayons. Rosa gave her 8 more crayons. How many crayons in all?

   _____ crayons

   Unit
   crayons

   Number model:

   _____

2. Use your calculator.

   Show 25.
   Broken key is ⑤.
   Show 2 ways:

   _____

   _____

3. Count back by 5s.

   45, 40, ____, ____, ____,

   ____, ____, 10, ____

   Can you keep going?

   0, ____, ____

4. Write <, >, or =.

   6 + 5 ____ 6 + 6

   8 + 3 ____ 12

   9 + 9 ____ 17

   4 + 9 ____ 8 + 5

5. Draw the hands to show 10:30.

6. Kyra has 2 dimes and 3 nickels in her left pocket. She has 1 quarter and 2 pennies in her right pocket. How much money does she have?

   _____

30 thirty

LESSON 2·4   **+9 Facts**

Write the sums.

1. 3
   + 9
   ___

2. 7
   + 9
   ___

3. ___ = 9 + 5

Unit

4. ___ = 2 + 9

5. 9
   + 4
   ___

6. 6
   + 9
   ___

7. 9
   + 8
   ___

8. 1 + 9 = ___

9. 0 + 9 = ___

10. 9
   + 9
   ___

11. 9 + ___ = 12

12. 15 = 9 + ___

Unit

13. ___ + 9 = 17

14. 13 = ___ + 9

Write a +9 number story.

_____

_____

_____

thirty-one **31**

Date _____ Time _____

LESSON 2·4 Math Boxes

1. Count by 6s.
Use your calculator.

_____, 12, _____, _____, 30

2. Fill in the missing numbers.

196	
	207

3. Solve.

Unit

16 = _____ + 1

_____ = 14 + 0

_____ + 9 = 9

_____ + 5 = 12

4. Show $0.88 in two ways.
Use Q, D, N, and P.

5. Room 10's Favorite Seasons

Season	Number of Children
Fall	////
Winter	//// //
Spring	////
Summer	//// ////

Which seasons have the same number of votes?

_____

6. 132 has… _____ hundreds

_____ tens

_____ ones

32 thirty-two

Date _____   Time _____

LESSON 2·5 Addition Facts

If you know a double, you know the 1-more and the 1-less sums.

Example:
If you know that 4 + 4 = 8,
You know   4 + 5 = 9,
And   4 + 3 = 7

1. 3 + 4 = _____   2. 8 + 7 = _____

3. _____ = 6 + 7   4.   6
                         + 5
                         ____

5.   8   6.   7   7.   7
    + 9   + 5   + 9
    ____   ____   ____

8. 5 + 8 = _____   9. _____ = 6 + 9   10. 8 + 6 = _____

Try This

11. 8 + 8 = _____   12. 12 + 12 = _____

    8 + 9 = _____   12 + 13 = _____

    8 + 7 = _____   12 + 11 = _____

13. 15 + 15 = _____   14. 14 + 12 = _____

    16 + 15 = _____   15 + 13 = _____

    14 + 15 = _____

thirty-three **33**

Date _____ Time _____

LESSON 2·5

Math Boxes

1. What shape is the cover of your math journal? Fill in the circle next to the best answer.

   Ⓐ rhombus

   Ⓑ rectangle

   Ⓒ triangle

   Ⓓ square

2. Circle the activity that takes about 1 second.

   Blinking your eyes.

   Writing your name.

   Reading a story.

3. Complete the fact family. Fill in the missing domino dots.

   8 + _____ = 15

   _____ + 8 = 15

   _____ = 15 − 8

   8 = 15 − _____

4. How likely is it that our school will have a fire drill today? Circle your answer.

   certain

   likely

   unlikely

   impossible

5. Put the numbers in order from smallest to largest. Circle the middle number.

   48, 44, 37, 54, 39

   _____, _____, _____, _____, _____

6. Draw hands to show 8:15.

34   thirty-four

LESSON 2·6 Domino Facts

For Problems 1 through 7, write 2 addition facts and 2 subtraction facts for each domino.

1. 4 + 2 = 6   2 + 4 = 6   6 − 2 = 4   6 − 4 = 2

2. ___ + ___   ___ + ___   ___ − ___   ___ − ___

3. ___ + ___   ___ + ___   ___ − ___   ___ − ___

4. ___ + ___   ___ + ___   ___ − ___   ___ − ___

5. ___ + ___   ___ + ___   ___ − ___   ___ − ___

6. ___ + ___   ___ + ___   ___ − ___   ___ − ___

Try This

7. ___ + ___   ___ + ___   ___ − ___   ___ − ___

8. Write one addition fact and one subtraction fact.

___ + ___   ___ − ___

thirty-five **35**

LESSON 2·6 Math Boxes

1. Fill in the missing numbers.

144		
	155	

2. What is the temperature? Fill in the circle next to the best answer.

- Ⓐ 55 degrees
- Ⓑ 62 degrees
- Ⓒ 52 degrees
- Ⓓ 56 degrees

3. Write the sums.

10 + 5 = _____

10 + 6 = _____

10 + 7 = _____

10 + 8 = _____

Unit

4. Write these numbers in order from smallest to largest. Begin with the smallest number.

133, 146, 129, 151

_____ , _____ , _____ , _____

5. Put an X on the digit in the tens place.

456

309

6. What time is it?

____ : ____

What time will it be in 15 minutes?

____ : ____

36 thirty-six

Date Time

LESSON 2·7 Subtraction Number Stories

Solve each problem.

1. Dajon has $11. He buys a book for $6. How much money does he have left?

    $____

2. Martin has 7 markers. Carlos has 4 markers. How many more markers does Martin have than Carlos?

    ____ markers

3. There are 11 girls on Tina's softball team. There are 13 girls on Lisa's team. How many more girls are on Lisa's team than on Tina's?

    ____ girls

4. Julia has 10 flowers. She gives 4 flowers to her sister. How many flowers does she have left?

    ____ flowers

5. Keisha has 8 chocolate cookies and 5 vanilla cookies. How many more chocolate cookies does she have than vanilla cookies?

    ____ chocolate cookies

6. Make up and solve your own subtraction story.

    _____

    _____

thirty-seven

Date   Time

LESSON 2·7 Math Boxes

1. Use your Pattern-Block Template. Draw a rhombus.

   There are _____ sides.

2. Circle the activity that takes about 1 minute.

   Brushing your teeth.

   Eating lunch.

   Playing a soccer game.

3. Write the fact family for this domino.

   ____ + ____ = ____
   ____ + ____ = ____
   ____ − ____ = ____
   ____ − ____ = ____

4. How likely is it that the school will serve lunch today? Circle your answer.

   impossible

   unlikely

   likely

   certain

5. Arrange the numbers in order from smallest to largest. Circle the middle number.

   98, 56, 143, 172, 81

   ____, ____, ____, ____, ____

6. What time is it?

   ____ : ____

   What time will it be in 30 minutes?

   ____ : ____

38   thirty-eight

Date _____ Time _____

LESSON 2·8 Using a Pan Balance and a Spring Scale

Weighing Things with a Pan Balance

1. Pick two objects. Which feels heavier?

   _____

2. Put one of these objects in the left pan of the pan balance.

3. Put the other object in the right pan.

4. Show what happened on one of the pan-balance pictures.

   ◆ Write the names of the objects on the pan-balance picture.

   ◆ Draw a circle around the pan with the heavier object.

5. Repeat with other pairs of objects.

scissors   pencil

Weighing Things with a Spring Scale

1. Which is heavier: 1 ounce or 1 pound? _____

2. How many ounces are in 1 pound? _____

3. Put objects in the plastic bag on the spring scale.

4. Weigh them. Try to get a total weight of about 1 pound.

5. List the objects in the bag that weigh a total of about 1 pound.

_____   _____   _____

_____   _____   _____

thirty-nine **39**

Date _____ Time _____

LESSON 2·8 Math Boxes

1. Fill in the missing numbers.

[T-shaped diagram with 13 in the top middle cell]

2. What temperature is it? Fill in the circle next to the best answer.

Ⓐ 61
Ⓑ 62
Ⓒ 64
Ⓓ 78

°F thermometer showing between 60 and 70

3. Write the sums.

10 + 7 = _____

10 + _____ = 12

_____ = 10 + 20

_____ = 10 + 41

Unit

4. Write these numbers in order from smallest to largest. Circle the smallest number and draw a box around the largest number.

243, 156, 326, 256

_____, _____, _____, _____

5. Put an X on the digit in the tens place in each number.

362   1,043

1,209   596

6. What time is it?

_____ : _____

What time will it be in 15 minutes?

_____ : _____

40   forty

LESSON 2·9

Name-Collection Boxes

1. Write 10 names in the 12 box.

12

2. Circle the names that DO NOT belong in the 9 box.

9
12 − 3   8 + 0
9 − 0   5 + 4 + 1
19 − 10   ⧸⧸⧸⧸ ///
x x x
15 − 7   x x x   1 less than 10
x x x
3 + 3 + 3   nine

3. Three names DO NOT belong in this box. Circle them. Write the name of the box on the tag.

9 + 3   12 − 8

3 + 3   ⧸⧸⧸⧸ //

x x x   5 + 3 − 2
x x x
                half
                a
10 − 4   dozen

4. Make up a name-collection box of your own.

forty-one **41**

| Date | Time |

LESSON 2·9 Pan-Balance Problems

Reminder: There are 16 ounces in 1 pound.

Some food items and their weights are shown below.

◆ Pretend you will put one or more items in each pan.

◆ Pick items that would make the balances tilt the way they are shown on journal page 43.

◆ Write the name of each item in the pan you put it in.

◆ Write the weight of each item below the pan you put it in.

Try to use a variety of food items.

Salad Dressing
1 ounce

Orange
8 ounces

Walnuts
3 ounces

Eggplant
15 ounces

Gummy Worms
4 ounces

Salt
1 pound

Lemon
6 ounces

Flour
2 pounds

Banana
6 ounces

Potatoes
5 pounds

42　forty-two

| Date | Time |

LESSON 2·9 Pan-Balance Problems *continued*

Example:

orange / gummy worms

Weight: __8 ounces__   Weight: __4 ounces__

1. Weight: _____   Weight: _____

2. Weight: _____   Weight: _____

3. Weight: _____   Weight: _____

4. Weight: _____   Weight: _____

forty-three 43

Date _____ Time _____

LESSON 2·9 Math Boxes

1. How much money?

   $ ___ . ____

2. Draw the hands to show the time school begins.

3. Write the fact family.

   (Triangle: 16, 9, 7, +,−)

   ____ + ____ = ____
   ____ + ____ = ____
   ____ − ____ = ____
   ____ − ____ = ____

4. Write the label and add 3 more names.

   20 − 4 |||| |||| |||| |

5. Use your Pattern-Block Template to draw a trapezoid.

6. How many books in all did Pedro read on Saturday and Sunday? Fill in the circle next to the best answer.

   Ⓐ 10   Ⓑ 5   Ⓒ 6   Ⓓ 11

44 forty-four

Date   Time

LESSON 2·10 Frames-and-Arrows Problems

1. Fill in the empty frames.

Rule: +8

[4] → [ ] → [ ] → [ ] → [ ]

2. Fill in the empty frames.

Rule: −4

(36) → ( ) → ( ) → ( ) → ( )

3. Fill in the empty frames.

Rule: +4

[ ] → [ ] → [15] → [ ] → [ ]

4. Fill in the arrow rule.

Rule:

(14) → (11) → (8) → (5) → (2)

Try This

5. Fill in the arrow rule and the empty frames.

Rule:

( ) → (20) → ( ) → (30) → ( )

forty-five **45**

LESSON 2·10 **Math Boxes**

1. Write 4 doubles facts that you know.

2. Complete the Fact Triangle and the fact family.

   12 +,− 3

   _____ = _____ + _____

   _____ + _____ = _____

   _____ − _____ = _____

   _____ = _____ − _____

3. Write the fact family for the domino.

   __11__ = _____ + _____

   _____ + _____ = __11__

   __11__ − _____ = _____

   __11__ − _____ = _____

4. Fill in the missing frames.

   Rule: +2

   5 ... 11

5. The total cost is 16¢. I pay with 2 dimes. How much change do I get? Fill in the circle next to the best answer.

   Ⓐ 36¢   Ⓑ 6¢
   Ⓒ 4¢   Ⓓ 20¢

6. Draw a rectangle around the digit in the tens place.

   3 4 9

   4 0 6

46 forty-six

LESSON 2·11 — "What's My Rule?"

In Problems 1–4, follow the rule. Fill in the missing numbers.

1. Rule: +6

in	out
2	8
3	9
5	
9	

2. Rule: −4

in	out
6	2
8	
10	
	5

3. Rule: +10

in	out
1	
5	15
	20
100	

4. Rule: −5

in	out
6	
	3
5	
12	

What is the rule? Write it in the box. Then fill in any missing numbers.

5. Rule: _____

in	out
6	13
1	8
3	
4	

6. Rule: _____

in	out
12	10
6	4
11	
	6

forty-seven 47

Date _____ Time _____

LESSON 2·11 Math Boxes

1. Show $1.50 three ways. Use Ⓠ, Ⓓ, and Ⓝ.

2. What time is it?

   ____ : ____

   What time will it be in 20 minutes?

   ____ : ____

3. Write the fact family.

   17
   +, −
   9   8

   ____ + ____ = ____

   ____ + ____ = ____

   ____ − ____ = ____

   ____ − ____ = ____

4. Write 6 names for 15.

   15

5. Draw a hexagon with your Pattern-Block Template.

   There are ____ sides.

6. How many cups of lemonade did Dee sell in the third hour? ____

48 forty-eight

LESSON 2·12   Math Boxes

1. Write the doubles fact.

   Number model:

   ____ + ____ = ____

2. Fill in the sum on the Fact Triangle. Write the fact family.

   8   5

   ____ + ____ = ____

   ____ + ____ = ____

   ____ − ____ = ____

   ____ − ____ = ____

3. Complete the fact family.

   9
   6   3

   ____ = ____ + ____

   ____ = ____ + ____

   ____ = ____ − ____

   ____ = ____ − ____

4. Fill in the frames.

   Rule
   +7

   ○ 14 ○ ○

   ○ 35

5. The total cost is 75¢. I pay with $1.00. How much change do I get?

   _____

6. Circle the digit in the hundreds place.

   128   972

   465   2,425

forty-nine 49

LESSON 2·13 Subtract 9 or 8

> **Reminder:** To find 18 − 9, think 18 − 10 + 1.
> To find 18 − 8, think 18 − 10 + 2.

1. Subtract. Use the −9 and −8 shortcuts.

   a. 13 − 9 = _____   b. 16 − 9 = _____   c. 14 − 8 = _____

   d. _____ = 12 − 8   e. _____ = 17 − 9   f. 12 − 9 = _____

   g. _____ = 13 − 8   h. 11 − 9 = _____   i. _____ = 15 − 8

   j.   15   k.   17   l.   11
      − 9   − 8   − 8

Try This

2. Find the differences.

   a. 43 − 9 = _____   b. 56 − 8 = _____   c. 65 − 9 = _____

   d. 37 − 8 = _____   e. 45 − 9 = _____   f. 53 − 8 = _____

3. Solve.

   a. 7 = _____ − 9   b. 6 = _____ − 8

50   fifty

Lesson 2·13 Math Boxes

1. How much money?

$_____ . _____

2. Show 8:50 P.M.

3. Find the turn-around facts.

3 + 4 = _____

4 + _____ = 7

8 + 5 = _____

5 + _____ = 13

4. Write the label and add 3 more names.

21 + 5
28 − 2
XXVI

5. Use your Pattern-Block Template to draw a trapezoid.

Circle the three sides that are the same length.

6. What day did Molly swim the most laps? Fill in the circle next to the best answer.

Days Molly Swam

Ⓐ Wednesday Ⓑ Sunday
Ⓒ Saturday   Ⓓ Tuesday

fifty-one 51

Date _____ Time _____

LESSON 2·14 Math Boxes

1. Selling Tickets for the School Fair

John
Andrea
Jason
0 5 10 15 20 25 30 35 40
Number of Tickets

Who sold the least amount of tickets? _____

2. Use the Pattern-Block Template to draw a hexagon.

3. Put an X on the digit in the tens place for each number below.

95

145

217

1,273

4. Draw the hands to show 7:45 A.M.

5. Write 6 names for $1.00.

$1.00

6. Fill in the missing frames.

Rule
−7

35 → ○ → 21
↓
7 → ○

Date	Time

LESSON 3·1 Place Value

Write the number for each group of base-10 blocks.

1. _____

2. _____

3. Write a number with …

    5 in the ones place,

    3 in the hundreds place, and

    2 in the tens place. _____

4. 506

    How many hundreds? _____

    How many tens? _____

    How many ones? _____

5. Marta wrote 24 to describe the number shown by these base-10 blocks:

    Is Marta right? Explain your answer.

    _____

    _____

    _____

    _____

fifty-three 53

Date _____ Time _____

LESSON 3·1 — Math Boxes

1. Fill in the frames. Use your calculator to count by 7s.

Rule: +7

7, 14, ___, ___, ___, ___

2. Write <, >, or =.

9 + 7 _____ 13

10 + 12 _____ 26

7 + 7 _____ 5 + 9

3. Solve.

Unit

7 + 8 = _____

70 + 80 = _____

700 + 800 = _____

7,000 + 8,000 = _____

4. How much money? Fill in the circle next to the best answer.

Ⓐ $21.40   Ⓑ $21.45

Ⓒ $2.45   Ⓓ $11.40

5. Mike had 7¢. He found a dime. How much money did he have in all? _____ ¢ Fill in the diagram and write a number model.

Start — Change → End

_____

6. Solve.

Unit

18 + 9 = _____

17 + 9 = _____

16 + 9 = _____

15 + 9 = _____

54 fifty-four

Date   Time

LESSON 3·2 Spinning for Money

Materials
- ☐ *Spinning for Money* Spinner (*Math Masters,* p. 472)
- ☐ pencil   ☐ large paper clip
- ☐ 7 pennies, 5 nickels, 5 dimes, 4 quarters, and one $1 bill for each player
- ☐ sheet of paper labeled "Bank"

Players 2, 3, or 4

Skill Exchange coins and dollar bills

Object of the Game To be first to exchange for a $1 bill

Directions

1. Each player puts 7 pennies, 5 nickels, 5 dimes, 4 quarters, and one $1 bill into the bank.

2. Players take turns spinning the *Spinning for Money* Spinner and taking the coins shown by the spinner from the bank.

3. Whenever possible, players exchange coins for a single coin or bill of the same value. For example, a player could exchange 5 pennies for a nickel, or 2 dimes and 1 nickel for a quarter.

4. The first player to exchange for a $1 bill wins.

Use a large paper clip and pencil to make a spinner.

fifty-five **55**

Date Time

LESSON 3·2 Fruit and Vegetables Stand Poster

Pear 13¢
Orange 18¢
Green Pepper 24¢
Melon slice 30¢
Plum 6¢
Banana 9¢
Apple 12¢
Tomato 20¢
Onion 7¢
Lettuce 45¢
Corn 15¢
Cabbage 40¢

56 fifty-six

Date	Time

LESSON 3·2 Buying Fruit and Vegetables

Select the fruit and vegetables from journal page 56 that you would like to buy. Write the name of each item.

Then draw the coins you could use to pay for each item. Write Ⓟ, Ⓝ, Ⓓ, or Ⓠ.

For Problems 3 and 4, write the total amount of money that you would spend.

I bought (Write the name.)	I paid (Draw coins.)	I paid (Draw coins another way.)
Example: one _orange_	Ⓓ Ⓝ Ⓟ Ⓟ Ⓟ	Ⓝ Ⓝ Ⓟ Ⓟ Ⓟ Ⓟ Ⓟ Ⓟ Ⓟ Ⓟ
1. one _____		
2. one _____		
3. one _____ and one _____		Total: _____

Try This

4. one _____, one _____, and one _____		Total: _____

fifty-seven **57**

LESSON 3·2 Math Boxes

1. Draw hands to show 4:30.

2. Write seven even numbers.

   _____ _____ _____

   _____ _____ _____

   _____

3. Put these numbers in order from smallest to largest and circle the middle number.

   23, 59, 49, 3, 159

   ____, ____, ____, ____, ____

4. Fill in the tally chart. Grade 1 sold 17 cupcakes at the bake sale and Grade 2 sold 13 cupcakes.

Number of Cupcakes Sold	
Grade 1	
Grade 2	

5. What is the temperature?

    _____

   Is it warm or cold?

   _____

6. A bag of pretzels costs 95¢. About how much money would you need to buy 3 bags of pretzels? Fill in the circle next to the best answer.

   Ⓐ $1.50   Ⓑ 95¢

   Ⓒ $3.00   Ⓓ $3.95

58 fifty-eight

Date _____ Time _____

LESSON 3·3 What Time Is It?

1. Write the time.

____ : ____   ____ : ____   ____ : ____   ____ : ____

2. Draw the hands to match the time.

8:00   3:30   7:45   9:15

3. Make up times of your own. Draw the hands to show each time. Write the time under each clock.

____ : ____   ____ : ____   ____ : ____

fifty-nine **59**

LESSON 3·3 Math Boxes

1. Use your calculator to count by 6s. Fill in the frames.

Rule: +6

36, ___, ___, ___, ___, ___, ___

2. Write <, >, or =.

6 + 7 _____ 15 − 4

5 + 8 _____ 8 + 5

18 − 9 _____ 5 + 4

3. Solve.

Unit

2 + 5 = _____

20 + 50 = _____

200 + 500 = _____

2,000 + 5,000 = _____

4. How much?

Q Q Q D D N P

_____

5. Chen had 30 postcards. He collected 17 more postcards. How many does he have in all?

_____

Fill in the diagram and write a number model.

Start | Change | End

6. Write the number that is 10 more.

19 _____

29 _____

39 _____

49 _____

60 sixty

Date _____ Time _____

LESSON 3·4

Build a Number

Your number	Show your number using base-10 blocks	Show your number another way using base-10 blocks

Date   Time

LESSON 3·4 Geoboard Dot Paper (7 × 7)

1.

2.

3.

4.

5.

6.

62 sixty-two

Date	Time

LESSON 3·4 Magic Squares

1. Add the numbers in each row. Add the numbers in each column. Add the numbers on each diagonal.

   Are the sums all the same? _____

2	9	4
7	5	3
6	1	8

   6 + 5 + 4 = ____

   2 + 9 + 4 = ____

   7 + 5 + 3 = ____

   6 + 1 + 8 = ____

   2 + 5 + 8 = ____

   ____ ____ ____

2. The sum of each row, column, and diagonal must be 15. Find the missing numbers. Write them in the blank boxes.

	7	
9		1
4		8

8		6
3		

sixty-three **63**

Date _____ Time _____

LESSON 3·4
Math Boxes

1. Draw hands to show 7:30.

2. Write odd or even next to each number.

5 _____

17 _____

98 _____

79 _____

3. Put these numbers in order from smallest to largest and circle the middle number.

12, 85, 77, 5, 71

_____, _____, _____, _____, _____

4.

Kickball Score	
Room 106	⧸⧸⧸⧸⧸ ⧸⧸⧸⧸⧸ //
Room 104	⧸⧸⧸⧸⧸ //

How many more points did Room 106 score than Room 104?

_____

5. What is the temperature?

_____

Is it warm or cold?

_____

6. A slice of pizza costs $2.12. A candy bar costs 52¢. If Jaleesa has $3.00, does she have enough money to buy the candy bar and the pizza?

_____

64   sixty-four

Date _____ Time _____

LESSON 3·5 Dollar Rummy

Materials
- ☐ *Dollar Rummy* cards (*Math Masters*, p. 454)
- ☐ scissors to cut out cards
- ☐ cards from *Math Masters*, p. 455 (optional)

Players 2

Skill Find complements of 100

Object of the Game To have more $1.00 pairs

Directions

1. Deal 2 *Dollar Rummy* cards to each player.

2. Put the rest of the deck facedown between the players.

3. Take turns. When it's your turn, take the top card from the deck. Lay it faceup on the table.

4. Look for two cards that add up to $1.00. Use any cards that are in your hand or faceup on the table.

5. If you find two cards that add up to $1.00, lay these two cards facedown in front of you.

6. When you can't find any more cards that add up to $1.00, it is the other player's turn.

7. The game ends when all of the cards have been used or when neither player can make a $1.00 pair.

8. The winner is the player with more $1.00 pairs.

sixty-five **65**

Date _____ Time _____

LESSON 3·5 **Pockets Data Table**

Count the pockets of children in your class.

Pockets	Children	
	Tallies	Number
0		
1		
2		
3		
4		
5		
6		
7		
8		
9		
10		
11		
12		
13 or more		

66 sixty-six

Date   Time

LESSON 3·5 Graphing Pockets Data

Draw a bar graph of the pockets data.

How Many Pockets?

Number of Children (y-axis: 0–10)

Number of Pockets (x-axis: 0, 1, 2, 3, 4, 5, 6, 7, 8, 9, 10, 11, 12, 13 or more)

sixty-seven **67**

Date	Time

LESSON 3·5 Math Boxes

1. Write 6 names in the 20-box.

20

2. How much money?

Q Q Q D P P

_____ ¢

3. Fill in the missing numbers.

Rule
−3

in	out
	4
10	
	6
	5

4. Solve.

5 + 3 = _____

30 + 50 = _____

3 + 6 = _____

60 + 30 = _____

Unit

5. I bought ice cream and a sandwich. Each cost 35¢. How much did I spend? _____ ¢

Fill in the diagram and write a number model.

Total	
Part	Part

6. What is the temperature? Fill in the circle next to the best answer.

Ⓐ 10°F   Ⓑ 55°F

Ⓒ 40°F   Ⓓ 50°F

68 sixty-eight

LESSON 3·6   Two-Rule Frames and Arrows

Fill in the frames. Use coins to help you.

1. **Rule** Add 5¢
   Rule Add 10¢

   5¢ → 10¢ ⇢ 20¢ → 25¢ ⇢ __ ← __ ⇠ __

2. **Rule** Add 10¢
   Rule Subtract 5¢

   __ → 20¢ ⇢ 15¢ → __ ⇢ __ ← __ ⇠ __

Try This

3. **Rule** Add 25¢
   Rule Add 5¢

   25¢ → 50¢ ⇢ __ → 80¢ ⇢ __ ← __ ⇠ __

sixty-nine **69**

LESSON 3·6 Two-Rule Frames and Arrows *continued*

4. Fill in the frames. Use coins to help you.

Rule: Add 10¢

3¢ → 13¢ → 12¢ → ___

Rule: Subtract 1¢

30¢ → ___ → ___

Fill in the frames and find the missing rules. Use coins to help you.

5. Rule: Add 2¢

1¢ → 3¢ → 4¢ → 6¢

Rule: ___

___ → ___ → 7¢

Try This

6. Rule: ___

$1.00 → $1.05 → ___ → $1.20

Rule: Add 10¢

$1.45 → ___ → ___

70 seventy

LESSON 3·6

Reading a Graph

Brothers and Sisters

Use the bar graph to answer each question.

1. How many children have 5 brothers and sisters? _____

2. How many children have 3 brothers and sisters? _____

3. What is the greatest number of brothers and sisters shown

   by the graph? _____ The least number? _____

4. What does the tallest bar in the graph show?

   _____

5. Suppose a new child came to this class. Predict how many

   brothers and sisters the child would have. _____

6. Explain why you think so.

   _____

   _____

seventy-one **71**

LESSON 3·6 Math Boxes

1. Write the amount. Fill in the circle next to the best answer.

Ⓝ Ⓝ Ⓠ Ⓟ Ⓓ Ⓓ

_____ ¢

Ⓐ 56¢
Ⓑ 55¢
Ⓒ 30¢
Ⓓ 46¢

MRB 88 89

2. Solve.

Unit

$70 + \underline{} = 100$

$100 = 20 + 30 + \underline{}$

$\underline{} = 50 + 40 + 10$

$100 = \underline{} + 10 + 60$

3. How many children have cats? _____

How many children have fish? _____

Children's Pets

(bar graph: Dogs, Cats, Birds, Fish; scale 0, 5, 10, 15, 20 Children)

4. Tomorrow I will get dressed for the day. (Circle one.)

Certain

Likely

Unlikely

Impossible

5. I had 17 tulips. I planted 20 more. How many do I have now? _____

Change

| Start | → | End |

Fill in the diagram and write a number model.

_____

6. Solve.

Unit

$\underline{} = 70 - 60$

$\underline{} = 88 - 80$

$63 - 20 = \underline{}$

$\underline{} = 92 - 30$

72 seventy-two

LESSON 3·7 Shopping at the Fruit and Vegetables Stand

Price per Item

pear	13¢	melon slice	30¢	lettuce	45¢
orange	18¢	apple	12¢	green pepper	24¢
banana	9¢	tomato	20¢	corn	15¢
plum	6¢	onion	7¢	cabbage	40¢

Complete the table.

I bought	I paid (Draw coins or $1 bill.)	I got in change
_____		_____ ¢
_____		_____ ¢
_____		_____ ¢

Try This

Buy 2 items. How much change from $1.00 will you get?

I bought	I paid	I got in change
_____ and _____	$1	_____ ¢

seventy-three **73**

Date _____ Time _____

LESSON 3·7 Math Boxes

1. Write 3 names for 50.

[50]

2. Write the fewest number of coins needed to make 67¢.

67¢ = _____ quarters

_____ dimes

_____ nickels

_____ pennies

3. Fill in the rule and the missing numbers.

Rule: _____

in	out
132	122
103	93
114	
205	

4. Solve.   Unit

6 + 8 = _____

80 + 60 = _____

7 + 4 = _____

40 + 70 = _____

5. 10 children ordered juice. 13 children ordered milk. How many children ordered drinks?

_____ children

Fill in the diagram. Write a number model.

Total
Part | Part

_____

6. What is the temperature?

_____

Would you wear a coat?

_____

74 seventy-four

Date   Time

LESSON 3·8 Making Change

Materials
- ☐ 2 nickels, 2 dimes, 2 quarters, and one $1 bill for each child
- ☐ 2 six-sided dice
- ☐ a cup, a small box, or a piece of paper for a bank

Number of children   2 or 3

Directions

1. Each person starts with 2 nickels, 2 dimes, 2 quarters, and one $1 bill. Take turns rolling the dice and finding the total number of dots that are faceup.

2. Use the chart to find out how much money to put in the bank. (There is no money in the bank at the beginning of the activity.)

Making Change Chart

Total for Dice Roll	2	3	4	5	6	7	8	9	10	11	12
Amount to Pay the Bank	10¢	15¢	20¢	25¢	30¢	35¢	40¢	45¢	50¢	55¢	60¢

3. Use your coins to pay the amount to the bank. You can get change from the bank.

4. Continue until someone doesn't have enough money left to pay the bank.

seventy-five

Date _____ Time _____

LESSON 3·8 Buying from a Vending Machine

1. The exact change light is on. You want to buy a carton of orange juice. Which coins will you put in? Draw Ⓝ, Ⓓ, and Ⓠ to show the coins.

2. The exact change light is off. You want to buy a carton of 2% milk. You don't have the exact change. Which coins or bills will you put in? Draw coins or a $1 bill.

    How much change will you get? _____

76 seventy-six

LESSON 3·8

Buying from a Vending Machine *continued*

3. The exact change light is on.

You buy:	Draw the coins you put in.
chocolate milk	
strawberry yogurt drink	
_____	

4. The exact change light is off.

You buy:	Draw the coins or the $1 bill you put in.	What is your change?
orange juice	Q Q Q	_____ ¢
chocolate milk	$1	_____ ¢
_____		_____ ¢
_____		_____ ¢
_____		_____ ¢

seventy-seven **77**

LESSON 3·8 Math Boxes

1. How much money?

Q Q Q Q D N
D N

_____ ¢ or $_____

2. Solve.

Unit

100 = 75 + ____

60 + ____ = 100

```
  95        90
+  5      + 10
```

3. How many children ate 2 scoops of ice cream?

_____

Scoops of Ice Cream

4. Tomorrow I will eat lunch. (Circle one.)

Certain

Likely

Unlikely

Impossible

5. How old will you be in 18 years?

_____ years old

Fill in the diagram and write the number model.

Start — Change → End

_____

6. Solve.

Unit

23 + 30 = ____

86 = ____ + 50

____ = 40 + 59

67 + 30 = ____

78 seventy-eight

LESSON 3·9   Math Boxes

1. A.M. temperature was 50°F.
   P.M. temperature was 68°F.
   What was the change? _____°F
   Fill in the diagram and write the number model.

   Start [ ]   Change →   End [ ]

   _____

2. Fill in the diagram and write a number model.

Total
Part 12	Part 18

   _____

3. I have $1.00. Can I buy two 45¢ ice cream bars?

   _____

4. What is the temperature?

   _____

5. Solve.   Unit
   22 + 10 = ____
   17 + 12 = ____

     45   100
   + 35   + 120

6. Solve.   Unit
   12 + 10 = ____
   13 + 10 = ____

     29   34
   +10   + 10

seventy-nine 79

LESSON 4·1
Fish Poster

Fish A
1 lb
12 in.

Fish B
3 lb
14 in.

Fish C
4 lb
18 in.

Fish D
5 lb
24 in.

Fish E
6 lb
24 in.

Fish F
8 lb
30 in.

Fish G
10 lb
30 in.

Fish H
14 lb
30 in.

Fish I
15 lb
30 in.

Fish J
24 lb
36 in.

Fish K
35 lb
42 in.

Fish L
100 lb
72 in.

80 eighty

Date   Time

LESSON 4·1   "Fishy" Stories

Use the information on journal page 80 for Problems 1–4.
Do the following for each number story:

◆ Write the numbers you know in the change diagram.
◆ Write "?" for the number you need to find.
◆ Answer the question.
◆ Write a number model.

1. Fish J swallows Fish B.

   How much does Fish J weigh now?

   Answer: _____ pounds

   Number model: _____

2. Fish K swallows Fish G.

   How much does Fish K weigh now?

   Answer: _____ pounds

   Number model: _____

3. Fish L swallows Fish F.

   How much does Fish L weigh now?

   Answer: _____ pounds

   Number model: _____

eighty-one   **81**

LESSON 4·1   **"Fishy" Stories** *continued*

Try This

4. Fish I swallows another fish.

   Fish I now weighs 20 pounds.

   Which fish did Fish I swallow?

   Answer: _____

   Number model: _____

   Start → Change → End

Write the missing number in each change diagram.

5. Start: 20   Change: +6   End: ___

6. Start: 30   Change: +50   End: ___

Try This

7. Start: 40   Change: ___   End: 45

8. Start: 9   Change: ___   End: 89

9. Start: ___   Change: +30   End: 39

10. Start: ___   Change: +3   End: 13

Date	Time

LESSON 4·1 Distances on a Number Grid

Use the number grid below. Find the distance from the first number to the second. Start at the first number and count the number of spaces moved to reach the second number.

1. 53 and 58 _____
2. 64 and 56 _____
3. 69 and 99 _____
4. 83 and 63 _____
5. 77 and 92 _____
6. 93 and 71 _____
7. 84 and 104 _____
8. 106 and 88 _____
9. 94 and 99 _____
10. 85 and 76 _____
11. 58 and 108 _____
12. 107 and 57 _____
13. 61 and 78 _____
14. 72 and 53 _____
15. 52 and 100 _____
16. 100 and 78 _____

51	52	53	54	55	56	57	58	59	60
61	62	63	64	65	66	67	68	69	70
71	72	73	74	75	76	77	78	79	80
81	82	83	84	85	86	87	88	89	90
91	92	93	94	95	96	97	98	99	100
101	102	103	104	105	106	107	108	109	110

eighty-three

Date _____ Time _____

LESSON 4·1 Math Boxes

1. Write the fact family.

   12
   +, −
   9   3

   ___ + ___ = ___
   ___ + ___ = ___
   ___ − ___ = ___
   ___ − ___ = ___

   MRB 27

2. What time is it? Circle the best answer.

   A. 6:30
   B. 6:25
   C. 6:27
   D. 5:30

3. Write the number that is 100 more.

   104 _____
   204 _____
   304 _____
   404 _____

4. How much money?

   $20 $20
   $10 $1
   Q Q Q D

   $ ____ . ____

5. Write <, >, or =.

   2,000 _____ 1,000
   2,500 _____ 3,500
   4,321 _____ 1,234
   5,674 _____ 5,674

   MRB 9

6. A baseball is shaped like a sphere. Name something else that is shaped like a sphere.

   _____

   MRB 57

84   eighty-four

Date _____ Time _____

LESSON 4·2 Parts-and-Total Number Stories

Lucy's Snack Bar Menu

Sandwiches		Drinks		Desserts	
Hamburger	65¢	Juice	45¢	Apple	15¢
Hot dog	45¢	Milk	35¢	Orange	25¢
Cheese	40¢	Soft drink	40¢	Banana	10¢
Peanut butter and jelly	35¢	Water	25¢	Cherry pie	40¢

For Problems 1–4, you are buying two items. Use the diagrams to record both the cost of each item and the total cost.

1. a soft drink and a banana

Total	
Part	Part

2. a hot dog and an apple

Total	
Part	Part

3. a soft drink and a slice of pie

Total	
Part	Part

4. a hamburger and juice

Total	
Part	Part

Try This

5. Jean buys milk and an orange. The cost is _____.

   Jean gives the cashier 3 quarters.

   How much change does she get? _____

eighty-five **85**

LESSON 4·2 Math Boxes

1. Solve.

Unit

8 + 7 = ____

80 + 70 = ____

800 + 700 = ____

8,000 + 7,000 = ____

2. A piece of candy costs 11¢. I pay with 15¢. How much change do I get? Circle the best answer.

A. 26¢   B. 4¢

C. 5¢   D. 6¢

3. Estimate.

Is 7 closer to 0 or closer to 10? ____

Is 53 closer to 50 or closer to 60? ____

Is 88 closer to 80 or closer to 90? ____

4. Circle names that belong.

$1.00

10 dimes

4 quarters   18 nickels

100 pennies

5 dimes   5 nickels

5. Circle the number sentences that are true.

9 + 7 = 7 + 9

8 − 5 = 5 − 8

6 + 5 = 5 + 6

6. Draw the other half of the shape and write the name of it.

_____

86   eighty-six

Date   Time

LESSON 4·3 Temperatures

Fahrenheit Thermometer °F
Celsius Thermometer °C

Water boils. 212°F, 100°C
Normal body temperature 98.6°F, 37°C
Room temperature 70°F, 21°C
Water freezes. 32°F, 0°C

1. Use a thermometer to measure and record the temperatures of the following:

   a. your classroom _____°F _____°C

   b. hot water from a faucet _____°F _____°C

   c. ice water _____°F _____°C

2. Which temperature makes more sense? Circle it.

   a. temperature in a classroom:

      40°F   or   70°F

   b. temperature of hot tea:

      100°F   or   180°F

   c. temperature of a person with a fever:

      100°F   or   100°C

   d. temperature on a good day for ice-skating outside:

      −10°C   or   10°C

eighty-seven **87**

Date _____ Time _____

LESSON 4·3 Math Boxes

1. Write the fact family.

14
+, −
7   7

_____ + _____ = _____

_____ − _____ = _____

2. What time is it?

_____

What time will it be in 1 hour?

_____

3. Count forward by 100s.

25, _____, 225, _____, _____, 525, 625

4. Show two ways to make 85¢. Use Ⓟ, Ⓝ, Ⓓ, and Ⓠ.

5. Write <, >, or =.

1,002 _____ 102

3,700 _____ 7,300

2,310 _____ 2,410

5,697 _____ 5,696

6. Your pencil tip is shaped like a cone. Name something else that is shaped like a cone.

_____

88   eighty-eight

Date _____ Time _____

LESSON 4·4 Parts-and-Total Number Stories

For each number story:
- Write the numbers you know in the parts-and-total diagram.
- Write "?" for the number you want to find.
- Answer the question. Remember to include the unit.
- Write a number model.

1. Jack rode his bike for 20 minutes on Monday. He rode it for 30 minutes on Tuesday. How many minutes did he ride his bike in all?

   Answer: _____
   (unit)

   Number model: _____

Total	
Part	Part

2. Two children collect stamps. One child has 40 stamps. The other child has 9 stamps. How many stamps do the two children have together?

   Answer: _____
   (unit)

   Number model: _____

Total	
Part	Part

3. 25 children take ballet class. 15 children take painting class. In all, how many children take the two classes?

   Answer: _____
   (unit)

   Number model: _____

Total	
Part	Part

eighty-nine 89

Date _____ Time _____

LESSON 4·4 Temperature Changes

Write the missing number in each End box.

Then fill in the End thermometer to show this number.

Unit
°F

1.
Start: 30
Change: +20
End: ___

2.
Start: 40
Change: −10
End: ___

3.
Start: 70
Change: −20
End: ___

4.
Start: 44
Change: −10
End: ___

90 ninety

Date _____ Time _____

LESSON 4·4 Temperature Changes *continued*

Fill in the missing numbers for each diagram.

Try This

5. Start [ ] — Change → End [ ]
_____

Start thermometer: 41°F
End thermometer: 52°F

6. Start [ ] — Change → End [ ]
_____

Start thermometer: 76°F
End thermometer: 60°F

7. Start [ ] — Change → End [ ]
_____

Start thermometer: 40°F
End thermometer: 53°F

8. Start [ ] — Change → End [ ]
_____

Start thermometer: 92°F
End thermometer: 80°F

ninety-one **91**

Date	Time

LESSON 4·4 Math Boxes

1. Solve.

Unit

_____ = 6 + 5

16 + 5 = _____

46 + 5 = _____

_____ = 76 + 5

86 + 5 = _____

2. LaVon has $1.00 and spends 73¢. How much change does she get?

_____

3. Estimate.

Is 49 closer to 40 or closer to 50? _____

Is 121 closer to 120 or closer to 130? _____

Is 214 closer to 210 or closer to 220? _____

4. Write 3 names for $2.00.

$2.00

5. Circle the number sentences that are true.

11 + 4 = 4 + 11

17 + 8 = 8 + 17

20 − 1 = 1 − 20

6. Draw the other half of the shape and write the name of it.

_____

Date _____ Time _____

LESSON 4·5 School Supply Store

Prices:
- pencil: 21¢
- pen: 45¢
- ruler: 37¢
- eraser: 28¢
- notebook: 69¢
- crayons: 76¢
- tape: 52¢

You have $1.00 to spend at the School Store.
Use estimation to answer each question.

Can you buy: Write *yes* or *no*.

1. a notebook and a pen? _____

2. a pen and a pencil? _____

3. a box of crayons and a roll of tape? _____

4. a pencil and a box of crayons? _____

5. 2 rolls of tape? _____

6. a pencil and 2 erasers? _____

7. You want to buy two of the same item.
   List items you could buy two of with $1.00.

   _____   _____

   _____   _____

8. How many pencils could you buy with $1.00? _____

ninety-three 93

Date _____ Time _____

LESSON 4·5 Math Boxes

1. I had $0.35. I spent $0.15. How much change do I have? Circle the best answer.

   A. $0.50   B. $0.15

   C. $0.45   D. $0.20

2. What is the temperature?

   _____ °C

   Is it warm or cold?

   _____

3. Write *even* or *odd* on the line.

   23 _____

   52 _____

   258 _____

   197 _____

4. Draw hands to show 5:15.

5. Fill in the frames.

   Rule: −2

   ○ ○ 90 ○ ○

6. Use your Pattern-Block Template. Draw a trapezoid.

   How many sides? ___ sides

94 ninety-four

Date   Time

LESSON 4·6 Shopping Activity

Do this activity with a partner.

Materials
- ☐ Shopping cards (*Math Masters*, p. 105)
- ☐ *Math Masters*, p. 433
- ☐ calculator
- ☐ play money for each partner: at least nine $1 bills and eight $10 bills; one $100 bill (optional)

Directions

1. Partners take turns being the Customer and the Clerk.
2. The Customer draws two cards and turns them over. These are the items the Customer is buying.
3. The Customer places the two cards on the parts-and-total diagram—one card in each Part box.
4. The Customer figures out the total cost of the two items without using a calculator.
5. The Customer counts out bills equal to the total cost. The Customer places the bills in the Total box on the parts-and-total diagram.
6. The Clerk uses a calculator to check that the Customer has figured the correct total cost.
7. Partners switch roles.
8. Play continues until all eight cards have been used.

Another Way to Do the Activity

Instead of counting out bills, the Customer says or writes the total. The Customer gives the Clerk a $100 bill to pay for the items. The Clerk must return the correct change.

ninety-five   **95**

Date _____ Time _____

LESSON 4·6   **Shopping Poster**

Telephone $46	Camera $43	CD Player $25	Calculator $17
Toaster $29	Iron $32	CD $14	Radio $38

96 ninety-six

Date _____ Time _____

LESSON 4·6 Shopping Problems

1. You buy a telephone and an iron. What is the total cost? $ _____

   Number Model: _____

Total

Part	Part
46	32

2. You buy a radio and a calculator. What is the total cost? $ _____

   Number Model: _____

Total

Part	Part
38	17

Solve each problem.

3. You bought two items. The total cost is exactly $60. What did you buy?

   _____

   _____

4. You bought two items. Together they cost the same as a telephone. What did you buy?

   _____

   _____

Try This

5. You bought three items. The total cost is exactly $60. What did you buy?

   _____

   _____

ninety-seven 97

Lesson 4·6 Math Boxes

1. 45 cents = 1 quarter and ____ dimes

   60 cents = 3 dimes and ____ nickels

2. A.M. temperature was 50°F.
   P.M. temperature is 68°F.
   What was the change? ____°F
   Fill in the diagram and write the number model.

   Start [ ] →Change→ End [ ]
   _____

3. 20 airplanes. 8 take off.
   How many stay? ____ airplanes stay. Fill in the diagram and write a number model.

   Start [ ] →Change→ End [ ]
   _____

4. Draw hands to show 8:15.

5. Measure the length of this line segment. Circle the best answer.
   _____

   A. about 7 cm

   B. about 6 cm

   C. about 8 cm

   D. about 10 cm

6. Draw a rectangle. Two sides are 4 cm long and two sides are 3 cm long.

98 ninety-eight

Date   Time

LESSON 4·7   Measuring Lengths with a Tape Measure

1. Measure the height from the top of your desk or table to the floor. Measure to the nearest inch.

   The height from my desk or table to the floor is about _____ inches.

2. Measure the height from the top of your chair to the floor. Measure to the nearest inch.

   The height from the top of my chair to the floor is about _____ inches.

3. Measure the width of your classroom door.

   The classroom door is about _____ inches wide.

ninety-nine   **99**

Date   Time

LESSON 4·7   Measuring Lengths with a Tape Measure *cont.*

4. Open your journal so it looks like the drawing below.

   a. Measure the long side and the short side to the nearest inch.

      The long side is about _____ inches.

      The short side is about _____ inches.

   b. Now measure your journal to the nearest centimeter.

      The long side is about _____ centimeters.

      The short side is about _____ centimeters.

100   one hundred

Date _____ Time _____

LESSON 4·7 — Tiling Surfaces with Shapes

Materials
- ☐ pattern blocks
- ☐ slates
- ☐ sheets of paper
- ☐ Pattern-Block Template
- ☐ scissors
- ☐ Everything Math Deck cards, if available

1. Pick one pattern-block shape. **Tile** a card by covering it with blocks of this shape.

   ◆ Lay the blocks flat on the card.

   ◆ Don't leave any spaces between blocks.

   ◆ Keep the blocks inside the edges of the card. There may be open spaces along the edges.

   Count the blocks on the card. If a space could be covered by more than half of a block, count the space as one block. Do not count spaces that could be covered by less than half of a block.

   Which pattern-block shape did you use?

   _____

   Number of blocks needed to tile the card:

   _____

   Trace the card. Use your Pattern-Block Template to draw the blocks you used to tile the card.

one hundred one **101**

Date　　　　　　　　　　　　　Time

LESSON 4·7   Tiling Surfaces with Shapes *continued*

2. Use Everything Math Deck cards to tile both a slate and a Pattern-Block Template. How many cards were needed to tile them?

Slate: _____ cards

Pattern-Block Template: _____ cards

3. Fold a sheet of paper into fourths. Cut the fourths apart. Use them to tile larger surfaces, such as a desktop.

Surface	Number of Fourths
_____	_____
_____	_____

Follow-Up

With a partner, find things in the classroom that are tiled or covered with patterns. Make a list. Be ready to share your findings.

Date _____ Time _____

LESSON 4·7 An Attribute Rule

Choose an Attribute Rule Card. Copy the rule below.

Rule: _____

Draw or describe all the attribute blocks that fit the rule.

Draw or describe all the attribute blocks that do *not* fit the rule.

These blocks fit the rule:

_____

_____

_____

_____

These blocks do *not* fit the rule:

_____

_____

_____

_____

one hundred three **103**

LESSON 4·7 Math Boxes

1. A peach costs 15¢. An apple costs 12¢. Show the coins needed to buy both. Use Ⓟ, Ⓝ, Ⓓ, and Ⓠ.

2. Show 67°F.

3. Write *even* or *odd* on the line.

    23 _____

    52 _____

    258 _____

    197 _____

4. Write the time.

    _____ : _____

5. Fill in the frames.

    +5   −1

    3   12

    20

6. Use your Pattern-Block Template. Draw a hexagon.

    There are _____ sides.

104 one hundred four

LESSON 4·8 Addition Practice

Write a number model to show the ballpark estimate. Solve the problem. Show your work in the workspaces.

Unit

1. Ballpark estimate:
   _____

   39
   + 26

2. Ballpark estimate:
   _____

   18
   + 45

3. Ballpark estimate:
   _____

   52
   + 28

4. Ballpark estimate:
   _____

   54
   + 79

5. Ballpark estimate:
   _____

   115
   + 32

Try This

6. Ballpark estimate:
   _____

   327
   + 146

Add. In each problem, use the first sum to help you find the other two sums.

7. 17 + 8 = ____

   17 + 8 + 25 = ____

   17 + 8 + 25 + 12 = ____

8. 15   15   15
   + 9   9   9
           + 6   6
                  + 22

9. 19 + 6 = ____

   19 + 6 + 5 = ____

   19 + 6 + 5 + 70 = ____

10. 24   24   24
    + 4   4   4
           + 7   7
                  + 35

one hundred five 105

LESSON 4·8 Math Boxes

1. How much?

   $10 $10 $5

   Q D D D D

   N N P P P

   $ _____

2. The temperature was 73°F. It got 13°F colder. What is the temperature now? _____ °F
   Fill in the diagram and write a number model.

   Start → Change → End

   _____

3. 25 books. Bought 15 more. How many now? _____ books

   Fill in the diagram and write a number model.

   Start → Change → End

   _____

4. What time is it?

   _____ : _____

   What time will it be in a half hour?

   _____ : _____

5. Draw a line segment 6 cm long. Underneath it, draw a line segment that is 2 cm longer.

6. Draw a rectangle. Two sides are 3 cm long and two sides are 5 cm long.

Date _____ Time _____

LESSON 4·9 — **Partial-Sums Addition Using Base-10 Blocks**

Draw base-10 blocks and write the number sentence to solve each problem.

1. **Example:**

   Unit: *books*

   ```
     23
   + 46
   ```

Tens	Ones
||	• • •
||||	• • • • • •

   60 + 9

   Answer: **69 books**

2. Unit: _____

   ```
     41
   + 35
   ```

Tens	Ones

   ___ + ___

   Answer: _____

3. Unit: _____

   ```
     67
   + 38
   ```

Tens	Ones

   ___ + ___

   Answer: _____

4. Unit: _____

   ```
     46
   + 36
   ```

Tens	Ones

   ___ + ___

   Answer: _____

Date _____ Time _____

LESSON 4·9 Addition Practice

Write a number model to show your ballpark estimate. Solve the problem. Show your work. Use the ballpark estimate to check whether your exact answer makes sense.

Unit

1. Ballpark estimate:

   _____

   ```
     59
   +  8
   ```

2. Ballpark estimate:

   _____

   ```
     67
   +  7
   ```

3. Ballpark estimate:

   _____

   ```
     47
   + 32
   ```

4. Ballpark estimate:

   _____

   ```
     58
   + 26
   ```

5. Ballpark estimate:

   _____

   ```
    122
   +  53
   ```

6. Ballpark estimate:

   _____

   ```
    136
   + 157
   ```

108 one hundred eight

Date _____ Time _____

LESSON 4·9 The Time of Day

For Problems 1–4, draw the hour hand and the minute hand to show the time.

1. Luz got up at 7:00.
   She had breakfast an hour later.
   Show the time when she had breakfast.

2. The second graders went on a field trip.
   They left school at 12:30.
   They got back 2 hours later.
   Show the time when they got back.

3. Tony left home at 8:15.
   It took him half an hour to get to school. Show the time when he arrived at school.

4. Ming finished reading a story at 10:30.
   It took her 15 minutes. Show the time when she started reading.

5. The clock shows when Bob went to bed.
   He went to sleep 15 minutes later.
   At what time did he go to sleep?

   _____ : _____

one hundred nine 109

LESSON 4·9 Math Boxes

1. I bought a radio for $67.00.
I paid with $100.00.
How much change did I get?

_____

2. What is the temperature?
Circle the best answer.

A. 34°F

B. 36°F

C. 32°F

D. 4°F

3. Write three even and three odd 2-digit numbers. Circle the even numbers.

_____   _____

_____   _____

_____   _____

4. What time is it?

____ : ____

What time will it be in 15 minutes?

____ : ____

5. Find the rules.

11   15   9   13

11   7

6. Use your Pattern-Block Template. Draw the triangles.

What is the difference between the triangles?

Date   Time

LESSON 4·10 Math Boxes

1. Draw a rectangle that has two sides that are 4 cm long and two sides that are 7 cm long.

2. Measure the line segment with the cm side of your ruler.

   _____

   How long is it? _____ cm

3. Use your Pattern-Block Template to draw a polygon.

   What is the name of the polygon you drew?
   _____

4. A 6-sided die is shaped like a cube. What is the shape of one of its sides? Circle the best answer.

   A. square

   B. triangle

   C. hexagon

   D. circle

5. Draw a line segment 2 cm long.

6. Use your template. Draw the other half of the shape and write the name of it.

one hundred eleven **111**

Lesson 5·1 Math Boxes

1. 843

There are _____ hundreds.

There are _____ tens.

There are _____ ones.

2. Read the tally chart. How many children in the class are 8 years old? _____

Class Ages

Age	Number of Children
7	𝍤 𝍤 /
8	𝍤 /
9	///

3. How many in all? Circle the best answer.

A 841   B 481

C 408   D 148

4. Fill in the missing numbers.

144

5. Draw a triangle. Make each side 4 cm long.

6. Graph this data. Jamar earned 3 stickers on Monday, 2 on Tuesday, and 4 on Wednesday.

Jamar's Sticker Graph

112   one hundred twelve

Date _____ Time _____

LESSON 5·2 Using Secret Codes to Draw Shapes

1. The codes show how to connect the points with line segments. Can you figure out how each code works? Discuss it with your partner.

Code:
A→E→D

Code:
E→A→C→B→D

Code:
E→A→C→D→B

2. Use each code. Draw line segments using your straightedge.

Code:
A→B→C→A

Code:
A→B→D→E→A

Code:
A→B→C→E→D→C→A

Code:
B→A→C→E→D→B

Code:
A→B→C→E→A

Code:
C→D→E→C→B→A→C

one hundred thirteen 113

LESSON 5·2

Math Boxes

1. Fill in the missing numbers.

Rule: +25

(circles: __, 50, __, __, 125, __)

2. Evan's Running Log

On which day did Evan run 8 miles? _____

3. Which number is an odd number less than 50? Circle the best answer.

A 50
B 37
C 26
D 55

4. Solve.

Unit

8 + 6 = _____

Write the turn-around fact.

_____

5. What time is it?

____ : ____

In 15 minutes it will be

____ : ____.

6. 12 red jelly beans. 16 green jelly beans. How many more green jelly beans? _____

Fill in the diagram and write a number model.

Quantity: 16
Quantity: 12
Difference: _____

114 one hundred fourteen

Date Time

LESSON 5·3 Parallel or Not Parallel?

These line segments are **parallel.**

These line segments are **not parallel.**

Quadrangles (Quadrilaterals)

These polygons are quadrangles (quadrilaterals).

square (ABCD)

rectangle (EFGH)

parallelogram (IJKL)

rhombus (MNPR)

trapezoid (STWU)

kite (XYZV)

one hundred fifteen   **115**

Date _____ Time _____

LESSON 5·3 **Parallel Line Segments**

Use a straightedge.

1. Draw line segments *AB* and *CD*.

   Are line segments *AB* and *CD* parallel?

   _____

2. Draw line segment *EF*.

   Are line segments *AB* and *EF* parallel?

   _____

3. Draw line segment *LM*.

4. Draw a line segment that is parallel to line segment *LM*. Label its endpoints *R* and *S*.

5. Draw a line segment that is NOT parallel to line segment *LM*. Label its endpoints *T* and *U*.

Try This

6. Draw a quadrangle that has NO parallel sides.

116   one hundred sixteen

LESSON 5·3 Parallel Line Segments continued

7. Draw a quadrangle in which opposite sides are parallel.

8. Draw a quadrangle in which all four sides are the same length.

What is another name for this shape?

_____

9. Draw a quadrangle in which 2 opposite sides are parallel and the other 2 opposite sides are NOT parallel.

one hundred seventeen 117

LESSON 5·3   "What's My Rule?" and Frames and Arrows

Find the rule. Then write the missing numbers in each table.

Unit
°F

1.

in	out
30	37
25	
	43
8	
	19

2.

in	out
50	30
	50
10	
20	
	42

Fill in the frames.

3. Rule +5   Rule +10

[ ] [ ] [ ] [30] [ ] [ ]

4. Rule +5   Rule +10

[ ] [25] [35] [ ] [ ] [ ]

118   one hundred eighteen

Lesson 5·3 Math Boxes

1. Use the digits 1, 3, and 5 to make:

the smallest number possible.

_____

the largest number possible.

_____

2. On which day did Ms. Forbes' class have the most recess minutes? _____

Ms. Forbes' Class Recess Minutes

Day	Number of Minutes
Wednesday	|||| |||| |||| ||||
Thursday	|||| |||| |||| |||| ||||
Friday	|||| |||| ||||

3. How many in all?

_____

4. Complete the number grid.

565

5. Draw a square. Make each side 3 cm long.

6. Graph this data.

On Tuesday it was 20°C, on Wednesday it was 35°C, and on Thursday it was 30°C.

3-Day Weather Report

Temperature (Celsius): 0°, 10°, 20°, 30°, 40°
Day of the Week: Tues, Wed, Thur

one hundred nineteen 119

LESSON 5·4 Polygons

Triangles

Quadrangles or Quadrilaterals

Pentagons

Hexagons

Heptagons

Octagons

These are NOT polygons.

Date _____   Time _____

Lesson 5·4 Math Boxes

1. Fill in the missing numbers.

Rule: −10¢

86¢ ○ ○ ○ ○
     ○ ○ ○

2. How many patients did Dr. Rios see on Tuesday? ____

Patient Log

(bar graph: M=15, Tu=10, W=20)
Patients Seen / Day of the Week

3. Write 5 even numbers greater than 50.

4. Solve.

9 + 3 = _____

Unit: cats

Write the turn-around fact.

_____

5. Draw the hands to show 6:45.

(clock face)

½ hour earlier is ___:___.

6. 35 butterflies. 10 flew away. How many butterflies are left?

_____ butterflies

Fill in the diagram and write a number model.

| Quantity |
| 35 |

| Quantity |
| 10 |

Difference

_____

one hundred twenty-one 121

LESSON 5·5 Math Boxes

1. How many cubes? _____

   Cross out 13 cubes.

   How many are left? _____

   Write the number model.

   ____ − ____ = ____

2. Fill in the missing numbers.

	112	
121		

3. Use your ruler. Draw $\overline{MS}$.

   $\overline{MS}$ is ____ cm long.

4. A triangle has _____ sides.

   A rhombus has _____ sides.

   A trapezoid has _____ sides.

   A hexagon has _____ sides.

5. How many dots are in this 4-by-6 array? Count by 4s.

   _____ dots

6. Write 3 numbers that add up to 20.

   ____ + ____ + ____ = ____

122 one hundred twenty-two

Date _____ Time _____

LESSON 5·6 Connecting Points

Draw a line segment between each pair of points.
Record how many line segments you drew.

Example:

A
C
B

3 points
__3__ line segments

1.

P
A
L
U

4 points
_____ line segments

2.

R
O
E
S
I

5 points
_____ line segments

Try This

3. Connect the points in order from 1 to 3. Use a straightedge.
Find and name 3 triangles.

_____

Try to name a fourth triangle.

_____

Color a 4-sided figure.

B
3
1
A
2
C

one hundred twenty-three **123**

Date   Time

LESSON 5·6 3-D Shapes Poster

Rectangular Prisms

Cylinders

Sphere

Cones

Pyramids

vertices

faces

edges

bases

curved surfaces

124 one hundred twenty-four

Date   Time

LESSON 5·6   What's the Shape?

Write the name of each shape.

1. _____

2. _____

3. _____

4. _____

5. _____

6. _____

7. _____

8. _____

9. _____

10. _____

one hundred twenty-five   **125**

Lesson 5·6 Math Boxes

1. Count by 1,000.

1,300; _____; _____; _____; _____; _____; _____; _____

2. Match quarters to ¢.

5 quarters — 150¢

6 quarters — 250¢

7 quarters — 125¢

10 quarters — 175¢

3. Ways to Get to School

bus
walk
car
bike

0 2 4 6 8 10 12 14
Number of Children

How many children walk to school? _____

4. Solve.   **Unit**

3 + 5 = _____

30 + 50 = _____

300 + 500 = _____

_____ = 6 + 8

_____ = 60 + 80

_____ = 600 + 800

5. Buy a model dinosaur for 37¢. Pay with 2 quarters.

How much change? _____

6. 4 children share 12 slices of pizza equally. How many slices does each child get? Draw a picture.

Each child gets _____ slices.

Date _____ Time _____

LESSON 5·7 Math Boxes

1.

How many cubes? _____

Cross out 14 cubes.

How many are left? _____

Write the number model.

_____ – _____ = _____

2. Complete the number grid.

442

3. Use your ruler. Draw line segment AB.

$\overline{AB}$ is _____ cm long.

4. This is a trapezoid. Put an X on the line segments that are parallel.

5. How many dots are in this 5-by-4 array? Count by 5s. Circle the best answer.

_____ dots in all.

A 19   B 20

C 25   D 15

6. Write 3 names for 15.

_____ + _____ + _____ = 15

15 = _____ + _____ + _____

_____ + 9 + _____ = 15

one hundred twenty-seven 127

Date   Time

LESSON 5·8 Symmetrical Shapes

Each picture below shows half of a shape on your Pattern-Block Template. Guess what the full shape is. Then use your template to draw the other half of the shape. Write the name of the shape.

Example:	1.	2.
rhombus		
3.	4.	5.
6.	7.	8.

128   one hundred twenty-eight

LESSON 5·8 Math Boxes

1. Count by 1,000.

2,600; _____; _____;

_____; _____; _____;

_____; _____

2. Write <, >, or =.

4 dimes _____ 50¢

3 quarters _____ 75¢

$1.00 _____ 11 dimes

3. Favorite Sport

soccer
basketball
football

0 1 2 3 4 5 6 7 8 9
Number of Children

How many more children like basketball than soccer?

_____

4. Solve.   Unit

18 − 9 = _____

180 − 90 = _____

1,800 − 900 = _____

5. Spend 68¢. Pay with $1.00. How much change?

_____

6. Share 19¢ equally among 5 children. How many cents does each child get? _____

How many cents are left over? Circle the best answer.

A 0¢   B 5¢

C 4¢   D 2¢

one hundred twenty-nine **129**

LESSON 5·9 **Math Boxes**

1. Draw a 7-by-9 array.

2. How many hot dogs were sold in the 2nd hour of the game? _____

   Hot Dogs Sold

3. Graph this data.

   On a nature walk Amy saw 5 snails, 2 turtles, and 6 caterpillars.

   Nature Walk

4. Three children share 18 pieces of paper equally.

   How many pieces of paper will each child get? _____

   How many pieces of paper will be left over? _____

5. How many dots are in this 4-by-4 array?

   _____

6. 28 soccer balls. 36 basketballs. How many more basketballs?

   _____ basketballs

   Fill in the diagram and write a number model.

Quantity
36

Quantity

   _____
   Difference

130   one hundred thirty

Date _____ Time _____

LESSON 6·1 Three Addends Directions

Materials ☐ number cards 0–20

Players 2

Directions

Skill Add three numbers.

Object of the Game
To find easy combinations when adding three numbers.

◆ Shuffle the cards. Place the deck number-side down.

◆ Turn over the top 3 cards. Each partner writes the 3 numbers.

◆ Add the numbers. Write a number model to show the order in which you added.

Unit

◆ Compare your answers with your partner's.

Example:

The cards 6, 5, and 14 are turned over. Gillian records the numbers. She adds 14 and 6 first and then adds 5. She records her number model and compares her answer with her partner's.

Numbers: __6__, __5__, __14__   Number model: __14__ + __6__ + __5__ = __25__

1. Numbers: ___, ___, ___
   Number model:
   ___ + ___ + ___ = ___

2. Numbers: ___, ___, ___
   Number model:
   ___ + ___ + ___ = ___

3. Numbers: ___, ___, ___
   Number model:
   ___ = ___ + ___ + ___

4. Numbers: ___, ___, ___
   Number model:
   ___ = ___ + ___ + ___

5. Numbers: ___, ___, ___
   Number model:
   ___ + ___ + ___ = ___

6. Numbers: ___, ___, ___
   Number model:
   ___ + ___ + ___ = ___

LESSON 6·1 Ballpark Estimates

Fill in the unit box. Then, for each problem:

◆ Make a ballpark estimate.

◆ Write a number model for your estimate. Then solve the problem. For Problems 1, 2, and 3 use the Partial-Sums Algorithm. For Problems 4–6, use any strategy you choose.

◆ Compare your estimate to your answer.

Unit

1. Ballpark estimate:

   _____

   29 + 7 = _____

2. Ballpark estimate:

   _____

   87 + 9 = _____

3. Ballpark estimate:

   _____

   37 + 42 = _____

4. Ballpark estimate:

   _____

   27 + 13 = _____

5. Ballpark estimate:

   _____

   38 + 46 = _____

6. Ballpark estimate:

   _____

   42 + 28 = _____

Date _____ Time _____

LESSON 6·1 Math Boxes

1. Which is least likely to happen? Choose the best answer.

- ⬭ A dog will have puppies today.
- ⬭ You will go to sleep tonight.
- ⬭ Chocolate milk will come out of the water fountain.
- ⬭ Someone will read you a book.

2. How long is this line segment?

_____

about _____ cm

3. Arlie harvested 12 bushels of corn and 19 bushels of tomatoes. How many bushels in all? _____ bushels

Fill in the diagram and write a number model.

Total	
Part	Part

_____

4. How many dots in this 2-by-8 array?

• • • • • • • •
• • • • • • • •

_____

5. Put these numbers in order from least to greatest and circle the middle number (the median).

109, 99, 129

6. Use your calculator. Start at 92. Count by 5s.

92, 97, _____, _____, _____,

_____, _____, _____, _____

one hundred thirty-three **133**

LESSON 6·2

Comparison Number Stories

For each number story:

♦ Write the numbers you know in the comparison diagram.

♦ Write ? for the number you want to find.

♦ Solve the problem.

♦ Write a number model.

1. Barb scored 27 points.
   Cindy scored 10 points.

   Barb scored _____ more points than Cindy.

   Number model: _____

   | Quantity |
   | Quantity |
   _____
   Difference

2. Frisky lives on the 16th floor.
   Fido lives on the 7th floor.

   Frisky lives _____ floors higher than Fido.

   Number model: _____

   | Quantity |
   | Quantity |
   _____
   Difference

3. Ida is 36 years old. Bob is 20 years old.

   Ida is _____ years older than Bob.

   Number model: _____

   | Quantity |
   | Quantity |
   _____
   Difference

LESSON 6·2   Comparison Number Stories *continued*

4. A jacket costs $75.
   Pants cost $20.

   The pants cost $ _____ less than the jacket.

   Number model: _____

   | Quantity |
   | Quantity |
   _____ Difference

Try This

5. Jack scored 13 points. Jack scored 6 more points than Eli.

   Eli scored _____ points.

   Number model: _____

   or _____

   | Quantity |
   | Quantity |
   _____ Difference

6. Billy is 16 years old. Paul is 6 years younger than Billy.

   Paul is _____ years old.

   Number model: _____

   or _____

   | Quantity |
   | Quantity |
   _____ Difference

7. Marcie is 56 inches tall.
   Nick is 70 inches tall.

   Marcie is _____ inches shorter than Nick.

   Number model: _____

   or _____

   | Quantity |
   | Quantity |
   _____ Difference

one hundred thirty-five   135

Date _____ Time _____

LESSON 6·2 Math Boxes

1. Fill in the frames.

+2   −5

( ) → (19) → ( ) → ( )

( ) ← ( ) ← ( ) ← ( )

MRB 98 99

2. Draw all the possible ways to show 30¢ using Q, D, N.

3. Write <, >, or =.

Unit

7 + 5 + 30 ____ 40

11 + 6 + 4 ____ 26

32 ____ 18 + 7 + 2

19 ____ 13 + 9 + 1

MRB 9

4. Draw a line segment that is about 5 cm long.

5. Write the time.

____ : ____

6. Halve.

2 ____   10 ____

4 ____   14 ____

8 ____   50 ____

136   one hundred thirty-six

LESSON 6·3

What Is Your Favorite Food?

1. Make tally marks to show the number of children who chose a favorite food in each group.

fruit/vegetables	bread/cereal/rice/pasta	dairy products	meat/poultry/fish/beans/eggs/nuts

2. Make a graph that shows how many children chose a favorite food in each group.

Favorite Foods of Children

Number of Children (y-axis: 0 to 30)

Food Groups (x-axis: fruit/vegetables, bread/cereal/rice/pasta, dairy products, meat/poultry/fish/beans/eggs/nuts)

one hundred thirty-seven **137**

LESSON 6·3

Comparing Fish

Fish Lengths

Fish B — 14 in.

Fish C — 18 in.

Fish D — 24 in.

Fish H — 30 in.

Fish K — 42 in.

Fish L — 72 in.

1. Fish C is _____ inches longer than Fish B.

Quantity
Fish C 18 in.

Quantity
Fish B 14 in.

   _____
   Difference

2. Fish H is _____ inches shorter than Fish K.

Quantity
Fish K _____

Quantity
Fish H _____

   _____
   Difference

138 one hundred thirty-eight

Date _____ Time _____

LESSON 6·3 Comparing Fish *continued*

3. Fish L is _____ inches longer than Fish H.

Quantity

Quantity
_____
Difference

4. Fish B is _____ inches shorter than Fish D.

Quantity

Quantity
_____
Difference

5. Fish H is 6 inches longer than _____.

Quantity

Quantity
_____
Difference

6. Fish L is 30 inches longer than _____.

Quantity

Quantity
_____
Difference

7. Fish C is 6 inches shorter than _____.

Quantity

Quantity
_____
Difference

8. Fish B is 16 inches shorter than _____.

Quantity

Quantity
_____
Difference

one hundred thirty-nine 139

Date _____ Time _____

LESSON 6·3 Math Boxes

1. Which of these is unlikely to happen? Circle the sentence.

   It is unlikely that...

   you will have lunch today.

   you will grow 5 inches while you sleep tonight.

   you will see someone you recognize.

2. Measure the line segment.

   _____

   about _____ in.

   about _____ cm

3. Denise picked 16 tulips and 4 daffodils. How many flowers did she pick in all?

   _____ flowers

   Fill in the diagram and write a number model.

Total	
Part	Part

   _____

4. How many dots are in this 3-by-5 array?

   • • • • •
   • • • • •
   • • • • •

   _____

5. List the numbers in order.
   5, 10, 4, 6, 4, 4, 6

   _____

   Which number occurs most often? _____

6. Continue.

   312, 314, 316, _____, _____, _____, _____, _____, _____

140 one hundred forty

Date _____ Time _____

LESSON 6·4 Addition and Subtraction Number Stories

Do the following for each problem:

◆ Choose one diagram from *Math Masters,* page 437.

◆ Fill in the numbers in the diagram. Write ? for the number you want to find. Find the answer. Write a number model.

◆ In Problems 1 and 2, write your own unit.

1. Colin has 20 _____.

   Fiona has 30 _____.

   How many _____ do they have in all?

   Colin and Fiona have _____

   _____ in all.
   (unit)

   Number model:

   _____

2. Alexi had 34 _____.

   He gave 12 _____ to Theo. How many

   _____ does Alexi have now?

   Alexi has _____

   _____ now.
   (unit)

   Number model:

   _____

3. Rushing Waters has 26 water slides. Last year, there were only 16 water slides. How many new slides are there this year?

   There are _____ new water slides.

   Number model:

   _____

4. The Loop Slide is 65 feet high. The Tower Slide is 45 feet high. How much shorter is the Tower Slide?

   It is _____ feet shorter.

   Number model:

   _____

one hundred forty-one **141**

LESSON 6·4 Ballpark Estimates

Fill in the unit box. Then, for each problem:

◆ Make a ballpark estimate before you add.

◆ Write a number model for your estimate.

◆ Use your calculator to solve the problem. Write your answer in the answer box.

◆ Compare your estimate to your answer.

Unit

1. Ballpark estimate:

_____

48 + 7

Answer

2. Ballpark estimate:

_____

63 + 9

Answer

3. Ballpark estimate:

_____

33 + 45

Answer

4. Ballpark estimate:

_____

23 + 14

Answer

5. Ballpark estimate:

_____

48 + 32

Answer

6. Ballpark estimate:

_____

45 + 38

Answer

142   one hundred forty-two

Date _____ Time _____

LESSON 6·4 Math Boxes

1. Fill in the frames.

−10   +2

( ) ( ) ( ) (139)
( ) ( )

2. Draw all the possible ways to show 35¢ using Q, D, N.

3. Write >, <, or =.

Unit

5 + 3 + 18 _____ 20

8 + 7 + 5 _____ 36

76 _____ 28 + 32 + 21

100 _____ 25 + 25 + 50

4. Which line segment is about 1 inch long? Choose the best answer.

○ ——————————
○ ——————
○ ————
○ ——

5.

1 hour earlier is _____:_____.

1 hour later is _____:_____.

6.

in	out
4	
8	
20	
	58

Rule
halve

one hundred forty-three 143

LESSON 6·5 Subtraction

Use base-10 blocks to help you subtract.

1.
Longs	Cubes
1	9
−	7

2.
Longs	Cubes
2	5
− 1	4

3.
Longs	Cubes
4	3
− 1	8

4.
Longs	Cubes
5	6
− 1	7

Use any strategy to solve.

5.   14
   − 6
   ―――

6.   38
   − 23
   ―――

7.   32
   − 19
   ―――

8.   64
   − 26
   ―――

Date _____ Time _____

LESSON 6·5 Math Boxes

1. Circle the one that is likely to happen.

It is likely that...

you will do a Math Box today.

you will fly like a bird.

an elephant will visit the classroom.

2. Measure the line segment.

_____

about _____ in.

about _____ cm

3. Kurtis scored 13 points in the first half of the game and a total of 24 points by the end. How many points did Kurtis score in the second half? _____ points

Total
Part	Part

Number model:

_____

4. How many dots are in this 7-by-9 array?

_____

5. Which number occurs most often? Choose the best answer.

8, 17, 9, 8, 10

◯ 9

◯ 17

◯ 10

◯ 8

6. Use your calculator. Count by 9s. Start at 76.

76, _____, _____, _____, _____, _____

What pattern do you see?

_____

_____

one hundred forty-five **145**

LESSON 6·6 **How Many Children Get *n* Things?**

Follow the directions on *Math Masters,* page 176 to fill in the table.

What is the total number of counters?	How many counters are in each group? (Roll a die to find out.)	How many groups are there?	How many counters are left over?

Date _____ Time _____

LESSON 6·6 Math Boxes

1. 639 has

   _____ hundreds

   _____ tens

   _____ ones

2. Draw the line of symmetry.

3. Share 1 dozen cookies equally among 5 children. Draw a picture.

   Each child gets _____ cookies.

   There are _____ cookies left over.

4. Use counters to make a 5-by-2 array. Draw the array.

   How many counters in all?

   _____ counters

5. The temperature is _____ °F.

6. Fill in the pattern.

   ◯ △ ◯ ___ ◯

   ___ ◯ ___

one hundred forty-seven 147

LESSON 6·7 Multiplication Stories

Solve each problem. Draw pictures or use counters to help.

Example: How many cans are in three 6-packs of juice?

```
///    ///    ///
///    ///    ///
 6      12     18
```

Answer: __18__ cans

---

1. Mr. Yung has 4 boxes of markers. There are 6 markers in each box. How many markers does he have in all?

Answer: _____ markers

2. Sandi has 3 bags of marbles. Each bag has 7 marbles in it. How many marbles does she have in all?

Answer: _____ marbles

3. Mrs. Jayne brought 5 packages of buns to the picnic. Each package had 6 buns in it. How many buns did she bring in all?

Answer: _____ buns

4. After the picnic, 5 boys each picked up 4 soft-drink cans to recycle. How many cans did the boys pick up all together?

Answer: _____ cans

Date _____ Time _____

LESSON 6·7 **Math Boxes**

1. Write the number.

   _____

2. Make ballpark estimates. Write a number model for each estimate.

   Unit

   32 + 59

   _____ + _____ = _____

   51 + 27

   _____ + _____ = _____

3. Measure the line segment.

   _____ cm

   Draw a line segment 3 cm shorter.

4.

   How many cubes? _____

   Cross out 23 cubes.

   How many are left? _____

   Write a number model.

   _____ − _____ = _____

5. Lauren's birthday is on the tenth day in the shortest month of the year. In what month is her birthday?

6. Double.

   Unit

   2 _____

   4 _____

   10 _____

   50 _____

one hundred forty-nine 149

Date _____ Time _____

LESSON 6·8 Array Number Stories

Array

o o o o o o o o o o
o o o o o o o o o o
o o o o o o o o o o
o o o o o o o o o o
o o o o o o o o o o
o o o o o o o o o o

Multiplication Diagram

rows	_____ per row	_____ in all

Number model: _____ × _____ = _____

Array

o o o o o o o o o o
o o o o o o o o o o
o o o o o o o o o o
o o o o o o o o o o
o o o o o o o o o o
o o o o o o o o o o

Multiplication Diagram

rows	_____ per row	_____ in all

Number model: _____ × _____ = _____

Array

o o o o o o o o o
o o o o o o o o o
o o o o o o o o o
o o o o o o o o o
o o o o o o o o o
o o o o o o o o o

Multiplication Diagram

rows	_____ per row	_____ in all

Number model: _____ × _____ = _____

Array

o o o o o o o o
o o o o o o o o
o o o o o o o o
o o o o o o o o
o o o o o o o o
o o o o o o o o

Multiplication Diagram

rows	_____ per row	_____ in all

Number model: _____ × _____ = _____

Date _____ Time _____

LESSON 6·8 Multiplication Number Stories

For each problem:
- ◆ Use Xs to show the array.
- ◆ Answer the question.
- ◆ Fill in the number model.

1. The marching band has 3 rows with 5 players in each row. How many players are in the band?

o o o o o o o o o o
o o o o o o o o o o
o o o o o o o o o o
o o o o o o o o o o
o o o o o o o o o o
o o o o o o o o o o

There are _____ players in the band.

_____ × _____ = _____

2. Mel folded his paper into 2 rows of 4 boxes each. How many boxes did he make?

o o o o o o o o o o
o o o o o o o o o o
o o o o o o o o o o
o o o o o o o o o o
o o o o o o o o o o

He made _____ boxes.

_____ × _____ = _____

3. The sheet has 5 rows of stamps. There are 5 stamps in each row. How many stamps are there?

o o o o o o o o o o
o o o o o o o o o o
o o o o o o o o o o
o o o o o o o o o o
o o o o o o o o o o
o o o o o o o o o o

There are _____ stamps in all.

_____ × _____ = _____

4. The orchard has 4 rows of trees. Each row has 8 trees. How many trees are there?

o o o o o o o o o o
o o o o o o o o o o
o o o o o o o o o o
o o o o o o o o o o
o o o o o o o o o o

There are _____ trees in the orchard.

_____ × _____ = _____

one hundred fifty-one 151

LESSON 6·8 **Math Boxes**

1. Choose the best answer. Be careful!

   6 tens

   3 ones

   8 hundreds

   ○ 638   ○ 836
   ○ 368   ○ 863

2. Draw the line of symmetry.

3. Use counters to solve.

   $14.00 is shared equally.

   Each child gets $5.00.

   How many children are sharing?

   _____ children

   How many dollars are left over?

   _____ dollars

4. This is a ____-by-____ array.

   How many dots in all?

   _____ dots

5. Find the differences.

   16°C and 28°C _____

   70°F and 57°F _____

   15°C and 43°C _____

6. What comes next?

152   one hundred fifty-two

LESSON 6·9 **Math Boxes**

1. How much?

   _____

2. Make a ballpark estimate.

   Write a number model for your estimate.

   49 + 51

   Number Model:

   _____

   Unit

3. Draw a line segment that is about 3 inches long.

   Now draw a line segment that is 1 inch shorter.

   How long is this line segment? Choose the best answer.

   ◯ 4 inches   ◯ 2 inches
   ◯ 1 inch   ◯ 5 inches

4. There are 22 first graders and 35 second graders. How many more second graders?

   _____ more

   Fill in the diagram and write a number model.

   Quantity

   Quantity

   _____

5. Write <, >, or =.

   1 week _____ 7 days

   48 hours _____ 1 day

   6 months _____ 1 year

6. Rule: double

in	out
2	
3	
	30
100	

one hundred fifty-three 153

Date	Time

LESSON 6·9 *Array Bingo* Directions

Materials
- ☐ 2 six-sided dice, 1 twelve-sided die, or an egg-carton number generator
- ☐ 9 cards labeled "A" cut from *Math Masters,* p. 450 for each player

Players   2–5

Skill   Recognize an array for a given number.

Object of the Game   Turn over a row, column, or diagonal of cards.

Directions

1. Each player arranges the 9 cards at random in a 3-by-3 array.

2. Players take turns. When it is your turn:

   Generate a number from 1 to 12, using the dice, die, or number generator. This number represents the total number of dots in an array.

   Look for the array card with that number of dots. Turn that card facedown.

3. The first player to have a row, column, or diagonal of facedown cards calls "Bingo!" and wins the game.

154   one hundred fifty-four

Date   **Time**

LESSON 6·9   *Array Bingo* Directions *continued*

Another Way to Play

Materials
- ☐ 1 twenty-sided die or number cards with one card for each of the numbers 1–20
- ☐ all 16 array cards from *Math Masters,* p. 450 for each player

Each player arranges his or her cards at random in a 4-by-4 array. Players generate numbers using the die or number cards. If you use number cards to generate numbers, do this:

- ◆ Shuffle the cards.
- ◆ Place them facedown on the table.
- ◆ Turn over the top card.
- ◆ If all 20 cards are turned over before someone calls "Bingo," reshuffle the deck and use it as before.

LESSON 6·10 Division Problems

Use counters or simple drawings to find the answers. Fill in the blanks.

1. **16 cents shared equally**

   by 2 people

   _____¢ per person

   _____¢ remaining

   by 3 people

   _____¢ per person

   _____¢ remaining

   by 4 people

   _____¢ per person

   _____¢ remaining

   by 5 people

   _____¢ per person

   _____¢ remaining

2. **25 cents shared equally**

   by 3 people

   _____¢ per person

   _____¢ remaining

   by 4 people

   _____¢ per person

   _____¢ remaining

   by 5 people

   _____¢ per person

   _____¢ remaining

   by 6 people

   _____¢ per person

   _____¢ remaining

3. **16 crayons, 6 crayons per box**

   How many boxes? _____   How many crayons remaining? _____

4. **24 eggs, 6 eggs in each cake**

   How many cakes? _____   How many eggs remaining? _____

LESSON 6·10 Math Boxes

1. Use the digits 3, 1, and 5.

   Write the smallest possible number.

   _____

   Write the largest possible number.

   _____

2. Do these objects have at least one line of symmetry?

   Yes or No?

   _____ _____

3. 18 cans of juice are shared by 5 people. Draw a picture.

   _____ cans per person

   _____ cans left over

4. 4 rows of 4 chairs. How many chairs in all? _____ chairs

   Draw an array to solve. Fill in the multiplication diagram.

rows	_____ per row	_____ in all

5. Find the differences.

   32°F and 53°F _____

   37°C and 19°C _____

   75°F and 93°F _____

6. Which figure does not belong? Choose the best answer.

   ○ circle with diagonal lines

   ○ empty circle

   ○ circle with vertical lines

   ○ circle with horizontal lines

one hundred fifty-seven **157**

Date _____ Time _____

LESSON 6·11 Math Boxes

1. Find the rule. Complete the table.

Rule

in	out
10	5
16	8
20	
	20

2. How much did Kristle weigh when she was 7 years old?

_____ pounds

Kristle's Weight Chart

3. Which number from this list occurs most often?

_____

3, 13, 23, 9, 14, 9

4. Double.

25¢ _____

50¢ _____

15¢ _____

75¢ _____

5. Draw the shapes that come next.

△ □ ○ ○ △

□ ___ ___

6. Count back by 10s.

220, 210, _____, _____,

_____, _____, _____,

_____, _____, _____,

_____, _____, _____

Table of Equivalencies

Weight

kilogram	1,000 g
pound	16 oz
ton	2,000 lb

1 ounce is about 30 g

<	is less than
>	is more than
=	is equal to
=	is the same as

Length

kilometer	1,000 m
meter	100 cm or 10 dm
decimeter	10 cm
centimeter	10 mm
foot	12 in.
yard	3 ft or 36 in.
mile	5,280 ft or 1,760 yd

10 cm is about 4 in.

Time

year	365 or 366 days
year	about 52 weeks
year	12 months
month	28, 29, 30, or 31 days
week	7 days
day	24 hours
hour	60 minutes
minute	60 seconds

Money

penny	1¢, or $0.01	P
nickel	5¢, or $0.05	N
dime	10¢, or $0.10	D
quarter	25¢, or $0.25	Q
dollar	100¢, or $1.00	$1

Capacity

1 pint = 2 cups
1 quart = 2 pints
1 gallon = 4 quarts
1 liter = 1,000 milliliters

Abbreviations

kilometers	km
meters	m
centimeters	cm
miles	mi
feet	ft
yards	yd
inches	in.
tons	T
pounds	lb
ounces	oz
kilograms	kg
grams	g
decimeters	dm
millimeters	mm
pints	pt
quarts	qt
gallons	gal
liters	L
milliliters	mL

Date _____ Time _____

LESSON 2·5   **+, − Fact Triangles 1**

Activity Sheet 1

LESSON 2·5 +, − **Fact Triangles 2**

Activity Sheet 2

LESSON 2·12 +, − Fact Triangles 3

Activity Sheet 3

Date		Time

LESSON 2·12 — +, − Fact Triangles 4

Activity Sheet 4